Watching Birds

An Introduction to Ornithology

by Roger F. Pasquier

Illustrated by
Margaret La Farge

Foreword by
Roger Tory Peterson

Houghton Mifflin Company · Boston

Library of Congress Cataloging in Publication Data

Pasquier, Roger F
Watching birds.

Bibliography: p.
Includes index.
1. Ornithology. 2. Bird watching. I. Title.
QL673.P37 598.2 77-1072
ISBN 0-395-25343-8 ISBN 0-395-29068-6 pbk.

Printed in the United States of America
V 10 9 8 7 6 5 4 3

In memory of

Herbert Dudley Hale

Preface

THE RECENT extraordinary growth in popularity of bird watching as a hobby and the wave of interest in exploring, protecting, and learning about the environment seem to me symptoms of a single desire: to become part of the natural world whose beauty and appeal lie in its independence of humanity. As our lives grow increasingly circumscribed by our own works, it becomes all the more fascinating to briefly step aside to watch, say, the birds that have landed in our yard for a day while on migration, knowing that they move entirely independent of our presence or concerns, part of a scheme more perfectly designed than the best of man's artistic or technical achievements. Proust explained the incongruity of trees flowering resplendently in a grimy suburb by saying they were traveling angels that had paused for a day; so too, the birds that pass through our yard or a city park are a reminder that there is a larger world than we initially perceive, one whose laws, manifested in the behavior of the birds, ultimately control our own.

Watching Birds: An Introduction to Ornithology is written for both the bird watcher and the environmentalist. The bird watcher, beginning or experienced, learns how to recognize different species, and often travels far to see them, but may never consider the reasons for their appearance, behavior, and distribution; this book gives him the tools to do so, by posing the questions he should ask himself while in the field, and providing some of the answers. It should also demonstrate that bird watching is much more fun when you stop to really *watch* the birds, hence the title of this book. The environmentalist or outdoors lover looks at the broad sweep of nature, but may lack a specific focus that visibly demonstrates how the general ecological principles he knows function in particular cases. Birds, as the most observable part of the natural world, abundantly demonstrate both the principles and the seeming exceptions; this book is written so that even if he has never watched birds as a hobby, the reader can fully understand the forces governing their existence, and can apply this understanding to the larger picture of all living things — it should make his explorations more fun as well.

Watching Birds is thus intended to unite the bird watcher's perception of specific details with the environmentalist's awareness of general truths. Although it explains some of the techniques of watching birds, this is not a book on identifying them — good ones already exist for that purpose — but a guide through the main topics of ornithology: how and why birds are studied, how they evolved, the physical adaptations that enable them to exploit such diverse roles in the environment, the significance of behavior in their daily lives and annual cycles, and how their distribution reflects broader zoogeographical principles. Following this survey, the realms of man and bird are integrated in chapters on conservation and on attracting and caring for birds, and the book concludes with a review of current major concerns of ornithology, reminding the reader that ornithology is unusual among the sciences in that the non-professional can still make significant contributions. A section of further reading lists books providing details on each of the subjects covered in the previous chapters.

This project began because Arthur Cooley, a science teacher at Bellport (N.Y.) High School, wanted a secondary-school-level ornithology text; he had taught ornithology at Bellport for three years to several classes of enthusiastic students per year, but had never found a satisfactory book or set of readings. I welcomed the challenge of writing such a book because for several years I had given informal field courses on birds and ecology to young children, teenagers, adults, and teachers, and had thought carefully what approach would most arouse their interest; writing it all down would enable me to include much more detail than I could give on a bird walk, and the information would be organized in a more logical sequence. When the project became a book for the general public, the style and format changed slightly, but I think it will still be useful to teachers and students.

Like many avid bird watchers, I started when very young. By the age of 7, I had begun looking at birds in New York City's Central Park, an oasis of green that to date has attracted at least 259 species and remains my favorite birding locale. For years my interest was simply in looking at and listing the birds I saw. Not until I began working at Great Gull Island, a remarkable research station in Long Island Sound managed by the American Museum of Natural History, where a long-term study of the breeding biology of Common and Roseate Terns was our daily occupation, did I become really interested in bird behavior and ecology. Recording the every move of selected tern families while watching from a blind, banding hundreds of chicks, weighing and measuring adults, and considering the ultimate question of how two such closely related species coexist in what seems like a uniform environment brought my interest to a new level. When I began leading bird walks for the Parks Council of New York, and then gave longer programs on birds and ecology at the Wave Hill Center for Environmental Studies, I tried to ask the participants the same questions I was asking myself — why do birds have certain colors and patterns? Why do they sing songs at specific pitches? Why do they have particular mating systems? I hoped the thinking process as well as the answers would give people a concept of natural selection and adaptation that they could apply when observing any animal or plant.

While writing this book I worked as a curatorial assistant in the Department of Ornithology at the American Museum and had the exciting opportunity to examine nearly every species — and most of the subspecies — of the world's birds. This gave me a new perspective on the birds I knew in life and their relationships to those from unfamiliar regions. I considered such questions as which forms within a family were closest to the ancestral type and which were more highly evolved, whether certain species were closely related or only similar by convergence, and which were in fact species and not just long-isolated races. Such thoughts made me believe that the underlying pleasure in looking at birds is the link they give not only to other places but also to the past and the future, reflecting the unending course of evolution and the most fundamental mysteries of life. I hope reading *Watching Birds* will arouse in others the same curiosity about birds and the role they play in the environment that my own experiences have given me.

As this is a book for the general public, the practice of citing references for specific facts, standard in scientific literature, has not been followed, because I felt it would clutter the text and because so many of the references would be to old issues of journals not accessible to most readers. The Further Reading section, however, lists many books containing lengthy expansions of the topics I have been able to treat only briefly, and the Appendix gives the names and addresses of the major ornithological journals from which I have drawn, so that those interested may subscribe.

The reader will notice that the English names of bird species are always capitalized, as in Short-tailed Albatross or Great Black-backed Gull, but when speaking generally, albatrosses and gulls are not. This practice prevents the confusion that might arise when referring to all albatrosses having short tails or to all black-backed gulls, as there are several in various parts of the world, or to common terns vs. Common Terns.

I am grateful to many people for the assistance and encouragement they have given me while I was working on *Watching Birds*. To be particularly thanked are Arthur P. Cooley, whose idea the project was and who used many chapters on his classes, Richard L. Plunkett and Henry Hope Reed, both helpful in getting the book started, Dennis Puleston, who examined many chapters of the manuscript and offered useful comments, and my colleagues in the Department of Ornithology at the American Museum, for answering numerous questions and suggesting sources of information, especially John Farrand, Jr., whose knowledgeable review of several chapters increased their accuracy. Editing by James Thompson of Houghton Mifflin has greatly improved the clarity and flow of my writing. The beautiful drawings by Margaret La Farge will, I am sure, add much to every reader's enjoyment and understanding of this book. Any errors of fact or interpretation are, however, entirely my responsibility.

Foreword

By Roger Tory Peterson

"BUT TELL ME sir, what good are birds?" Even in this day of environmental awareness there are those who ask that question. They are biologically illiterate of course. In exasperation I can only reply, "Birds are alive, aren't they?"

Professor George Gaylord Simpson, speaking specifically of penguins, was more articulate when he answered, "That depends on what you mean by good. If you mean 'good to eat,' you are perhaps being stupid. If you mean 'good to hunt,' you are surely being vicious. If you mean 'good as it is good in itself to be a living creature enjoying life,' you are not being crass, stupid, or vicious. I agree with you." *

As I have emphasized elsewhere, and it bears repeating — birds are far more than robins, cardinals, and wrens to brighten the garden, ducks and quail to fill the sportsman's bag, or rare shorebirds and warblers to be ticked off by the birder. They are indicators of the health of our planet — a sort of "ecological litmus paper." Because of their furious pace of living they reflect changes in the environment — the environment we all share. If they were removed, the balance of nature as we know it would be drastically upset.

I, for one, would find the world quite desolate if the birds were eliminated. They have been my obsession since boyhood, an obsession from which I have never freed myself. But the shuddering thought of a world without birds is shared by a great many other people. Much of the shock effect of Rachel Carson's book on pesticides was inherent in its title, *Silent Spring.*

To the readers of this book I need not spell out all the reasons why watching birds is well worth our time. I would risk triteness if I

* *Penguins. Past and Present, Here and There.* Yale University Press, 1976.

extolled their beauty, their voices, their incisive ways. The point I wish to make here, and it is precisely what Roger Pasquier had in mind when he wrote this book, is that identifying birds is not all there is to bird watching. It is phase number one. It can become a most compelling game of course. Some devotees never tire of it; they become sharper, tick off larger and larger lists, and spot more and more rarities.

So that I will not be accused of academic snobbery, let me hasten to say that I do not belittle this heady activity. As the one who launched the *Field Guides* I have had a lot to do with the burgeoning interest in birding and the passion to make lists. Any game or sport is worthwhile in its own right, and particularly if it is pursued to the championship level.

But to be champion inevitably demands a great deal of time and the wherewithal to travel. If the hard-core birder is to pass much beyond the 600 mark in North America, he must criss-cross the continent from the U.S.-Mexican border to Alaska, and from the Atlantic seaboard to California. To pass 700, as Joseph Taylor has done, he must have a network of spies to alert him about those birds he still "needs." To try for the world record he must compete with Stuart Keith who already has seen more than 5000 of the planet's 8700 species.

Many have tried to explain the pastime of bird watching. Joseph Hickey once commented: "By some, it is regarded as a mild paralysis of the central nervous system, which can be cured only by rising at dawn and sitting in a bog. Others regard it as a harmless occupation of children, into which maiden aunts may sometimes relapse."

Hickey wrote that bit of whimsy after he had bridged the gap from the bird lister to the bird watcher. He had gone through the usual cycle with his field glass and little white check-lists. The law of diminishing returns had set in and his zealous interest lapsed. He still went on a few field trips but it looked as though another ardent follower of the cult had been lost when Dr. Ernst Mayr, then a young ornithologist at the American Museum in New York, took an interest in Hickey and opened his eyes to the untapped possibilities of his hobby. He induced him not to identify a bird and then pass on quickly to the next one, but to slow down and inquire into its way of life. Hickey changed from a lister to a full-fledged bird watcher and to pay off his debt wrote his classic book, *A Guide to Bird Watching.** It demonstrated that bird study could last a lifetime; it pointed out what the amateur could do to add his bit to the science of ornithology. Written more than thirty years ago, it was the lineal ancestor of this new book by Roger Pasquier. Much has happened in the field of ornithology and in amateur bird watching during those three decades. Roger Pasquier represents the new wave of young ornithologists, both amateur and professional, who are evolving as Hickey did in the atmosphere of the Linnaean Society of New York and the American Museum of Natural History. His book is an excellent up-to-date extension of the *Field Guides.* Although birds do not change, bird watchers do.

Knowing more about birds should not lessen our feeling of wonder. Henry Beston, marvelling at the aerial maneuvers of the constellations of shorebirds on the beaches of Cape Cod, commented, "We need another wiser and perhaps more mystical concept of animals. Remote from universal nature, and living by complicated artifice, man in civilization surveys the creature through the glass of his knowledge and sees thereby a feather magnified and the whole image in distortion. We patronize them for their incompleteness, for their tragic fate of having taken form so far below ourselves. And therein we err, and greatly err. For the animal shall not be measured by man. In a world older and more complete than ours they move finished and complete, gifted with extensions of the senses we have lost or never attained, living by voices we shall never hear. They are not brethren, they are not underlings; they are other nations, caught with ourselves in the net of life and time, fellow prisoners of the splendour and travail of the earth." †

* *A Guide to Bird Watching.* Oxford, 1943; Dover, 1975.
† *The Outermost House.* Rinehart, 1947.

Contents

Watching Birds

An Introduction to Ornithology

Chapter One

These silhouettes show the distinctive flight postures that simplify identification of a crane, a heron or egret, and an ibis.

Watching Birds

STUDYING BIRDS offers rewards that can affect all parts of your life. As a hobby, bird watching has few equals, combining as it does the pleasures of hunting, because you must carefully stalk a wild creature to observe it closely, with those of collecting, since you may keep lists of the birds you see. The search for birds will take you to places of breathtaking beauty you might not otherwise visit, and areas near home you may have previously overlooked or found uninteresting. You will meet other bird watchers who can share their knowledge and experience with you — and you'll probably discover other mutual interests as well. The more you look at birds, the more the rest of the natural world they inhabit will arouse your curiosity, and the more you will appreciate how all living things, including man, are interconnected and dependent on the same environment for their well-being.

Of all the groups in the animal kingdom, birds are the most easily studied. There is no time of year or place in the world where you cannot find them; they are common in our biggest cities, on distant oceanic islands, and in the most seemingly inhospitable wildernesses. Ranging in size from the two-and-a-half-inch Cuban Bee Hummingbird to herons and cranes more than four feet tall, they are by and large active all year, colorful, noisy, and — compared with most other animals — fearless and easy to see. Approximately 650 of the world's 8600 to 9000 bird species (experts disagree on how many forms should be considered full species) are regularly found in North America; another 50 wander here as vagrants from other parts of the world, and new ones are sighted regularly. In the past few years birds that are not only new to our area but entirely new to science have been discovered as nearby as Puerto Rico and Hawaii.

Today, the study of birds is especially relevant to environmental concerns. Like the canaries coal miners took into deep mine shafts to warn them if oxygen was running low, wild birds serve as excellent indicators of the quality of our environment. Birds inhabit every environment in North America, and they feed at every level of the food chain pyramid — that is, from plant material, at the bottom, to animals that have eaten other animals that have eaten plants, at the top. The presence, variety, and numbers of birds in any habitat tells us a lot about the quality of that environment.

Before attempting any advanced studies, of course, you must learn to identify birds in the field. This in itself is a study, but always fun. The only equipment you need is a field guide to the birds of your area and a pair of binoculars. At first you may feel overwhelmed by the number of birds, but they will quickly fall into natural groupings based on family and habitat. Some birds are colorful, distinctively marked, and impossible to mistake, but others are not so easy — don't expect to identify every bird you see on your first field trips. Certain species and families of birds are famous for being hard to tell apart — the little sandpipers nicknamed "peep," flycatchers in the group called *Empidonax,* and the "confusing fall warblers"; even experts have trouble, and some don't even try to distinguish those species unless they have seen or heard them very well. The more often you watch birds, the easier the difficult ones will become, and the pleasure of identifying new species will outweigh any frustration.

After every trip into the field, you should write a few things down, including the date and localities, a list of all species seen, how many (at least approximately) of each, notes on what they were doing (singing, nest building, feeding young, migrating, etc.), weather conditions, and anything else you might think important. Even if this information seems trivial at first, it will grow in value as you build up a collection of notes — in your second year you will be able to compare dates of arrival and departure of migrants, how the nesting schedule may be affected by weather conditions, whether populations have changed, and other useful data. Over time, your records may show natural or manmade alterations in habitat in a particular area and the changes they cause in birdlife. You may live in an area where such studies have never been done; if so, your work could be especially useful should an ecologically distinctive area you have watched be threatened with alteration — your data would provide valuable information for assessing the impact of any proposed changes. In an area already well known, you may be able to compare your findings with descriptions from earlier periods and evaluate changes in bird populations over long stretches of time, telling you which birds increase, decrease, or are unaffected by different types of land use.

The handiest books to carry with you are Peterson's *Field Guide to the Birds* (for the eastern half of North America), his *Field Guide to Western Birds,* or *Birds of North America* (covering the whole continent) by Robbins, Bruun, and Zim. Each of these books illustrates all the birds found in the area it covers, and describes identifying marks, habitat, voice, and range; many people prefer to consult more than one guide. It is useful to study these books before you go bird watching, so you will know how the birds are arranged in the book and what to expect in your area during each season. Some species will be found year round, and others only in summer or winter, or as migrants passing through in spring and fall.

Birds are not often easy to see with the naked eye, as they may be far away or moving quickly, but for most kinds of birds watching, a pair of binoculars (7x35 and 8x40 are the commonest powers, and easiest to use) is all the help you need. For looking at waterfowl or shorebirds across lakes or mud flats, a higher power telescope is useful, but since you won't need it nearly so often as binoculars, get binoculars and learn to use them first. Binoculars with a central focus are easiest. You needn't get an expensive pair; as long as you have no trouble seeing through them, can focus to various distances easily, and don't drop them on the ground or in water very often, a cheap pair should serve your purposes.

When beginning, many people have difficulty locating birds, especially moving ones, through binoculars. If you start by focusing on objects that will not move away, you will soon know instinctively what adjustments you have to make to see clearly at any distance. When you see a bird, keep your eye on it while raising the binoculars, so that you are always looking in the same place; if you take your eye off the bird, you may not find it again through binoculars, because the instrument's field of vision is much narrower than that of the naked eye.

For many kinds of bird watching the best time to be in the field is early morning; during migration and the nesting season, small birds are most active and vocal then, and so are easier to spot. Birds seem to rest around midday or early afternoon and are more active again in the last hours of light. During the winter, however, when birds must feed all through the day and are not singing, there is little or no advantage to early morning bird watching. Water birds like ducks, geese, gulls, and herons are active all through the day, but soaring hawks, vultures, and eagles are rarely seen early in the morning because they usually wait until the sun has warmed the air sufficiently to create columns of rising warm air, called thermals, on which they can ride without effort. Unless you can get to the mud flats where shorebirds feed, they are most easily seen at high tide, when water covers many feeding areas and they concentrate on the few flats or sandbars above water.

Wherever you live, you will find birds throughout the year, but you can probably find the greatest variety during spring and fall migrations, when species nesting or wintering elsewhere pass through your region. For many people, migration is the most exciting time for bird watching, with each day offering the surprise of new arrivals, the possibilities of vagrants off their usual routes, and "wave days" when the trees are crowded with unusual concentrations of warblers, vireos, tanagers, orioles, and thrushes. In many parts of North America spring migration begins in February and may run, with different species on different schedules, into June; by early July some shorebirds appear in the northern states, already heading south, while ducks may not arrive in these same areas until December — thus there are few months in the year when some transient birds may not be in your area.

Where to look for birds depends on what you want to see: every bird has certain environmental requirements, a particular habitat to which it is tied, its niche. To find most ducks, geese, herons, gulls, and terns, for example, you must go where there is water, since their food requirements (fish, crustacea, aquatic plants, and insects) are found nowhere else. Most plovers and sandpipers feed on mud flats or, where these don't exist, in wet fields of short grass. Sometimes birds learn to exploit modern ecological alternatives — gulls thrive at garbage dumps and migrant shorebirds are found at sewage beds. Land birds each have specific niches when nesting, but are less particular on migration, when treetop birds may be seen in shrubbery or on the ground, and ground birds are sometimes found in trees. During migration, the best place to look for concentrations of land birds is often near fresh water — a stream, woodland pond, marsh, or lake edged with vegetation. Some birds are attracted to water by the insects, often available there earlier in spring and later in fall than elsewhere, many come to drink and bathe, and others may be lured by the warmth from the sun, which can strike all levels of a tree exposed by the water's edge. A similar reason may help explain why more birds are found at the sunny edges of woodlands than within a stand of trees. A woodland edge is also the edge of an adjoining habitat, so birds from two habitats may overlap here; the edges of fields, telephone rights-of-way through woods, and, in fact, the area where any two habitats join, called an ecotone, are especially good places to look for birds.

Bird watching is not restricted to "the country"; most cities have excellent birding localities. In spring and fall, city parks and well planted cemeteries are especially productive, as migrants encountering otherwise inhospitable urban areas concentrate in the green spaces; in fact, your chances of seeing the rarer migrants and large numbers of the commoner species are better here than in the larger stretches of countryside outside the city. In parks as in more wild environments, the best areas will be near water and at the edges of lawn or field.

Finding birds, especially the small land birds, often involves listening as much as looking, since birds at the tops of trees or in thick shrubbery may sing or call much more often than they make themselves visible. Learning how to pinpoint the direction from which the noise comes will help you to get a look at the bird, and learning the songs and call notes themselves is an enormous help. When you know the sounds made by the common birds you will no longer have to track each one down; recognizing the notes of rarer species will help you spot them more quickly. Since nearly every species has a distinctive song and many have distinctive calls, they are not hard to learn, only requiring some time to hear them in the field. You can also listen to records of bird calls and songs. A set of records or tape cassettes with the sounds of each species, one at a time, matching the Peterson guides, is available, as well as several records of bird choruses as you actually hear them in the field, with several species singing at once.

Once you have located a bird, identification is primarily a process of elimination; in addition to clues from sounds, many other things will help you to narrow the possibilities. Consider first the habitat in which you find the bird — you don't expect to see a sandpiper in a tree, a hawk walking on a mud flat, a meadowlark in the forest, or a woodpecker swimming. You will quickly learn each bird's usual activity and habitat. Many birds can easily be eliminated by range and season. If you put a mark in your guide at the descriptions and pictures of all species found in your area, you will save yourself a lot of time; although vagrants do appear, you will not encounter many of the birds that live in other sections of the continent. Similarly, know which birds to expect and which are unlikely at each season of the year. If, for example, you see a small gray and white bird with white outer tail feathers feeding on seeds in a snow-covered field, chances are it's a junco, but a bird of this description catching insects in the treetops of a southern forest in June is likely to be a gnatcatcher; in this way you can identify many birds on the basis of range, season, habitat, and activity, with only a minimal description of what they look like.

When looking at a bird, examine first its general outline and size. The shapes of body parts such as bill, neck, wings, and tail will give you hints. Herons, ibises, and cranes, for example, all have long necks, but in flight only herons hold their head back at the shoulders, with the resulting short-necked appearance. The ultimate source of bird identification is color and markings. Has the bird solid patches of color, streaks (stripes running lengthwise), bars (stripes across), or a combination? Has it got a stripe on the side of

Barring and striping. The adult Broad-winged Hawk (left) has bars across the underparts and tail, while the immature (right) has striped underparts.

Eye ring, wing bars, and eye stripe. The Solitary Vireo (left) has a prominent white eye ring and wing bars; the Red-eyed Vireo has an eye stripe and plain wings.

its face, running through the eye (an "eye stripe"), a ring around the eye ("eye ring"), or "wing bars," one or two stripes on the wing near the shoulder? These are some of the things to look for. Some birds are naturally more difficult to identify than others, but with a field guide and practice you can master most of the birds found in your area within a year. Identifying birds on your own is one of the greatest satisfactions of bird study, and essential to learning more about them.

Chapter Two

A Lark Bunting caught in a mist net. The bird is unharmed and can be removed easily for banding and examination.

How and Why Birds are Studied

THERE ARE many reasons for studying birds. Beside the fun of simply watching them, you will get much enjoyment from learning why birds look and act as they do. When you know what enables a bird to fly, why its plumage is of certain colors and patterns, or where it is going, the pleasure of seeing it in the field is much greater.

The scientific study of birds is called ornithology, from two Greek words: *ornis*, bird, and *logia*, a suffix form of *legein*, to speak. The English word "bird" comes to us only slightly changed from the Old English *brid*, a young bird, which may ultimately derive from Old English *bredan*, to cherish or keep warm: the origin of "bird" suggests a long and affectionate interest in birds by man.

The History of Bird Study

The long history of man's interest in birds, evidenced by many appearances in art, literature, and history, also adds to the fascination of studying birds today. Undoubtedly the first men studied birds mainly to learn how they were most easily caught, but the early appearance of birds in art suggests they played a role in folklore as well. The first known representations are cave sketches of a crane or heron, a stork, and an owl done between 16,000 and 15,000 B.C. in southern France and Spain, by the Aurignacian people. Between 9000 and 6000 B.C., the Magdalenians, the last Old Stone Age people of southern France, carved birds on reindeer antlers and engraved them on pebbles. In the early Neolithic period, between 6000 and 4000 B.C., inhabitants of southern Spain drew on cave walls at least a dozen recognizable types of birds, all of which are still found in Spain.

The world's first civilizations, the Sumerians and Egyptians, used carvings and paintings of birds to decorate residences, temples, and tombs as far back as 3000 B.C. The Egyptians considered some birds the embodiment of various deities; falcons, ibises, and vultures were especially esteemed. Ancient Egyptian paintings include some species no longer found in Egypt, indicating changes in bird distribution over the last 5000 years that would otherwise have been unknown.

In ancient China, birds appeared on decorative objects and in legends, and bird carvings exist from the Mayan and Aztec civilizations of Central and South America, where certain birds were also worshipped. References to birds are numerous in the Bible, but only about thirty species can be identified from the Biblical descriptions.

To our knowledge, Aristotle (384–322 B.C.) was the first person to look at birds scientifically, describing what he considered to be 170 varieties and discussing their physiology, reproduction, and ecology. Aristotle was also the first to make a systematic classification of birds; based on observation and dissection, he divided birds into eight groups. A few other Greeks and Romans observed and wrote about birds in a scientific manner, but for more than a thousand years, until the thirteenth century, little was added except fanciful myths and anecdotes. In the thirteenth century curiosity about the natural world revived, and from then on we have many accounts of birds as they relate to hunting, agriculture, cooking, travel, and other human pursuits; birds also began then to appear more frequently in literature — especially poetry — and in painting, tapestry, and sculpture. Lists and accounts grew longer and more detailed in the centuries after printing was invented. The first modern scientific classification of birds was done by an Englishman, Francis Willughby (1635–72), a century before the binomial system now in use was devised by Linnaeus (1707–78), a Swede. Explorations of all corners of the earth in the last two hundred years have made us aware of nearly all the world's birds, and scientific study has grown to cover all aspects of bird biology: classification, anatomy, physiology, evolution, behavior, migration, and the other topics that we will explore.

Birds have played a role in American history from the very beginning: Columbus's crew, doubting they were anywhere near land, was threatening mutiny, when migrating land birds were seen flying over the sea; Columbus changed his course and followed the birds to land — had he sailed in a month when birds were not migrating over the Atlantic, history might well have been different. The records of every early colony in North America contain amazed comments on the abundance of bird life; turkeys, waterfowl, and Passenger Pigeons must have sustained many early settlements through difficult periods. Specimens of North American birds were sent back to Europe to be examined and named, and by the eighteenth century men were exploring the continent specifically to study its wildlife. Mark Catesby (1679?–1749), who traveled through the wilderness between the Carolinas and Florida, described many new species of birds in his *The Natural History of Carolina, Florida and the Bahama Islands;* Alexander Wilson (1766–1813) attempted to include all that was known of American birds in his nine-volume *American Ornithology.* The magnificent paintings of John James Audubon (1785–1851) in *The Birds of America,* a collection of 435 life-sized hand-colored lithographs, illustrated every bird known from North America that Audubon had either observed or received as a specimen. The life histories that accompany the illustrations, together with Wilson's contributions and the work of other scientists in America, brought American ornithology close to the professional level the subject had in Europe; since that time, Americans have participated in every aspect of ornithological research.

Reasons for Studying Birds Today

Birds have been studied longer and in greater detail than any other form of animal life, but this should not suggest that everything important about them has been discovered and analyzed. The wealth of information already accumulated allows students to probe ever deeper for the smallest details and the most underlying principles of life, so that today we ask not only how birds sing, but also why each Song Sparrow sings an individual combination of notes and song phrases whose pattern can only be understood by a computer analysis of audio-spectrograms; answers to the most specific questions may help solve the larger ones.

Recently developed technologies such as radar, computers, and advanced sound recording and interpreting instruments have answered questions and opened new fields of study unavailable to earlier researchers, but much can still be learned by a person using his own eyes, ears, and head — ornithology is unusual among sciences in that it offers many opportunities for nonprofessionals to contribute. Mrs. Margaret M. Nice, a housewife raising five children, spent twelve years watching the Song Sparrows in her backyard and wrote one of the most detailed investigations ever published, *Studies in the Life History of the Song Sparrow,* which has served as a model for other workers. The *Life Histories of North American Birds,* a 23-volume standard reference covering nearly every species found in North America, was compiled and in large part written by a former businessman, Arthur C. Bent. Infinitely many other contributions have been made by people who have less time to devote to the subject; you can learn a lot during a few minutes' observation every day or a few weeks of the year.

If you are interested in the basic principles of ecology, you will find that birds are often the ideal way to see these principles in operation. Investigations of food chain pyramids, habitat partitioning, succession of plant and animal communities, evolution, speciation, distribution, and other topics relevant to all living things frequently use birds as examples, because they are often the easiest part of the environment to study. From your own observations of birds you may be able to draw conclusions about other forms of life.

Many aspects of bird study are of direct relevance to man. Investigations of food habits, for example, have changed the public's opinion of certain birds now known to be valuable predators of insects, rodents, and other animals man considers "pests." Only recently has the public become aware of this service birds perform. Many birds, particularly hawks, owls, and some fish eaters, were persecuted because they were believed to prey upon animals of value to man, but most people now realize that hawks and owls eliminate many rodents that might damage the farmers' crops, and fish eaters rarely take "sport" fish. That hawks and owls are beneficial was only proved to farmers and hunters after studies of stomach contents showed an almost entire absence of chickens and game birds, but very high percentages of crop-damaging rodents. Similarly, herons, kingfishers, and cormorants were shown to feed heavily on fish of no commercial value. Birds take only the natural surplus in the population of their prey, but in certain cases, especially artificial situations of prey abundance as found on a farm, birds may reduce numbers of the undesirable insect or rodent to a level where it no longer significantly damages a crop.

Birds are also excellent indicators of environmental quality. Declines in bird populations have alerted scientists and the public to changes in the environment; the United States ban on DDT and some other pesticides was achieved largely by demonstrating the severe effects that repeated spraying had on birds that ate the insects and fishes contaminated by the poisons. Human beings, larger and more long-lived, will not show the harmful effects as quickly, but our bodies may well have concentrations of the same pesticides.

The environmental impact statements required by law for many types of land development often use birds as indicators of the pre-alteration quality and the effects that changes might have. Birds found in specialized habitats like wetlands, frequently subject to environmental review before development can take place, are useful indicators of vegetation types, water quality, and food production. An area attractive to birds may also have positive values for people — wetlands that feed and shelter birds are also flood barriers, important nurseries for fish and shellfish, and places for recreation. Valuable, sometimes ecologically unique, land areas have been preserved because their alteration or destruction would endanger bird populations.

Many people are directly concerned with the life histories of ducks, geese, quail, and other game birds, which must be known to assure that only a replaceable number are taken each year by hunters and that there are adequate refuges throughout the birds' ranges. Food requirements, migration routes, tolerance of disturbance, nesting success, and other factors are all important in deciding when and where man can pursue these birds without damage to their population.

In some parts of the world, birds contribute directly to the economy as tourist attractions, sought for their beauty or their numbers: the Gannets of Bonaventure Island, Quebec, the herons and egrets in the Florida Everglades, the flamingos of Lake Nakuru, in Kenya, and even the Cliff Swallows at San Juan Capistrano, California are known to people all over the world. Maintaining these beneficial relationships with man requires more than just preserving the site itself; life histories must also be well known to assure that no actions of man elsewhere inadvertently reduce their numbers.

Other birds are of interest to man as sources of food. As noted, settlers of North America and other uncultivated parts of the world depended on wild birds as a source of meat. Some birds were used to replenish ship supplies; this practice entirely eliminated the Dodo of Mauritius Island in the Indian Ocean, illustrated on page 240, by the late seventeenth century. Today, wild birds and their eggs are used for food in several parts of the world, but investigations of their population and reproductive rate have determined how many can be taken safely; only abundant species are utilized, such as ducks in Iceland, shearwaters in Australia, and eggs of the Lapwing, a plover, in Holland and Germany, and in all of these cases the collecting is strictly regulated to assure the birds' continued abundance.

Bird feathers have long had economic importance for man. In America until about 1910 the plumes of egrets, herons, and birds-of-paradise and the entire bodies of small birds were used to decorate women's hats. Strong conservation laws now prevent such uses, and species once nearly exterminated for their plumes, like egrets, are again common. Down from ducks, geese, and swans is still used for clothing and bedding. The best down comes from eider ducks, which grow it specially to insulate their nests. In Iceland and Norway the down is gathered on a large scale, but the birds are carefully protected.

Guano, the accumulation of excrement from fish-eating birds, is another important economic product requiring study of bird habits to assure its continued availability. Colonies of penguins, terns, cormorants, and other seabirds on islands in dry climates, where rain does not wash the excrement away, produce enormous quantities of this extremely valuable fertilizer. The greatest source is the colonies of the Guanay Cormorant, on islands off Peru, where, after an early period of over-exploitation, only the annual yield is now harvested.

Birds are also studied for their roles in both spreading and preventing human diseases. Encephalitis, a sometimes deadly virus disease transmitted by ticks and mosquitoes, can be carried long distances by birds. Infected birds usually develop immunity quickly; their role in transporting the disease is probably infrequent, but still not clear. Histoplasmosis is a lung disease caught by inhaling the fungus spores in bird droppings; it is only a problem to people working in poultry yards or bird roosts, and 95 percent of those exposed soon become immune. Pigeons and parrots can transmit forms of virus pneumonia, but again only rarely. On the preventive side, chicks are sometimes used in making antimalarial vaccines, and wild birds, of course, consume many insects carrying diseases like yellow fever and malaria. The Sacred Ibis of Egypt is well known as a consumer of the snails which are a secondary host of the schistosome parasite, a blood fluke, but recent persecution and alteration of the ibis' habitat, reducing its population, have led to a rise in the disease. The transmission or control of other diseases may also be affected by birds in subtle ways not yet known; this awaits further study.

How Birds are Studied

Life history and ecological studies often involve work in laboratories and museums, but almost always start in the field. An interesting project to begin with is a local bird survey, indicating the movements and habitat requirements of the species present in each season. Investigating what birds feed on, how they nest — alone or in colonies — and where, territory size if there is one, and how they interact with other birds and animals will give clues to the basic ecological principles that govern all bird behavior.

For a life history study, choose a species that can be easily observed and watch it through the nesting cycle or the entire year if possible. Note how it sets up and defends a territory, finds a mate, interacts with others of its species, feeds, builds a nest, and cares for young; if you watch a few pairs of the same species you may find that certain individuals handle various aspects differently — while birds generally conform to set patterns of behavior, there is always room for individual variation, so that you will find some longer singers, more aggressive defenders, better foragers, or more diligent incubators than others. Watching a bird feeder or bath will reveal other variations in behavior among individuals and between species.

More advanced field studies often involve marking individual birds, usually by putting a numbered aluminum band around one leg; when correctly applied, the band does not interfere with the bird's activities or increase its weight. Audubon was the first person to band birds out of scientific curiosity: in 1803 he tied silver threads to the legs of young Eastern Phoebes in Pennsylvania; the next year he found that some of them returned to the same place to nest. Banding began on a large scale in 1903, at a German station, Rossitten, on the Baltic Sea. Before the purpose of banding birds to track their movements was well known, people who found banded birds had little idea of what they signified. When, shortly after banding began at Rossitten, a banded gull was found in France, people thought the band had been put on by a sailor on board a sinking ship named

Rossitten; when a Bulgarian shot a Spotted Eagle with band number 1285, the local press carried the story of a bird over 600 years old. Today in North America all bands are issued by and marked for return to the U.S. Fish and Wildlife Service, and a special permit is required to use bands. Birds are usually caught in harmless metal traps or with fine silk mesh nets, called mist nets, made of dark thread and erected on poles. Birds don't see the nets and after flying into them can be easily removed without injury; mist nets are the most efficient way to catch large numbers of small birds. Recoveries, dead or alive, of banded birds have given us information otherwise unobtainable on migration routes, travel speeds, and longevity.

For life history studies, birds can also be given combinations of colored plastic bands that permit the observer to identify individual members of a family or colony under study without continually having to retrap them; this is especially useful in species where the sexes cannot be told apart by external characteristics or where known individuals are being studied over several years.

Studies of the living bird use laboratory work to reveal things undiscoverable in a natural environment. All studies of anatomy and physiology, of course, require dissecting and examining internal parts to see how they function or compare with those of related species. How each organ serves a bird can sometimes be learned only by modifying or removing it: that owls can locate prey by sound was discovered when owls kept in a totally dark room, where eyes were no help, were found to have little trouble catching mice released there; with plugged ears, the owls could no longer locate the mice.

Other laboratory studies investigate what parts of bird behavior are inherited and which are "learned" by association of young birds with their parents or other birds. Chapter 7 describes how some birds reared in complete isolation never develop proper songs; after exposure to singing members of their species, they too may sing an appropriate song, while those exposed to the songs of other species may attempt these songs rather than their own. As discussed in Chapter 10, very important information on timing and orientation of migration has come from birds kept in laboratories: when exposed to certain amounts of light or patterns of stars, some will exhibit "migratory restlessness," hopping and beating against the side of the cage that faces the direction in which they wish to travel during spring or fall; artificial manipulation of these stimulants has shown how each factor affects migration impulses. As these examples show, much laboratory work is concerned with how birds perceive and react to various phenomena; by isolating the several phenomena to which birds are exposed all at once in the wild, the effect of each can be measured.

Museum collections of bird skins, skeletons, and eggs are vital to many kinds of bird study, including, at times, field work. In any collection you will find a label on each specimen, giving its scientific name and, if known, place and date of collection, age, and sex. Study skins have everything except the bones of the feet, wings and skull removed, and are filled with cotton to lie on their back in a standard position that facilitates comparative measurements. Preserved from dampness and feather-eating insects, a skin may last indefinitely; it is not uncommon to see a skin dating from the mid-nineteenth century next to a much more recent specimen, and in just as good condition. Large collections have specimens from every family of the world's birds, and almost every species; good smaller collections at colleges, nature centers, and natural history museums will have at least a representative sampling of all birds found locally. The largest collection in North America, with over 900,000 skins, is in New York City at the American Museum of Natural History. There are nineteen other institutions with at least 34,000 skins each, and several hundred more throughout the United States and Canada with a few hundred to a few thousand skins, as well as skeletons and eggs. Skins allow you to examine details of plumage, bill or foot structure, and the comparative measurements of body parts. Things impossible to see at one time in the field, even with birds in the hand, such as the sequence of plumages from hatching to maturity, or the patterns of molt, can be studied easily. Comparisons can also be made of individuals from different parts of a bird's range; in many species, populations from different areas differ slightly in size and color. Our ideas of how birds are related to one another comes in considerable measure from careful examinations of large series of skins and skeletons. Whether birds exposed to pesticides such as DDT laid eggs with thinner than normal shells was in part established by comparisons with eggs in museums, collected before pesticides were used. As some birds become increasingly scarce or extinct, museum specimens may be the only way we have to study them; however, no bird population was ever depleted by a museum collector — the 4 million specimens in North American museums consist mostly of commoner species, and the total is still less than the number of Mallard ducks shot *annually* in the United States alone.

Another major aspect of bird study involves familiarity with the literature and knowing where to find information on particular topics. No one can possibly keep up with all the books and magazines published on birds, but most fairly general bird books have listings of more specialized publications; checking bibliographies in some basic bird books in your library will save a lot of work when you're looking for information on a narrower topic.

Magazine articles describing the results of experiments and field work can be quite technical, but they may suggest research topics or techniques you can adapt to the birds around you and the time at your disposal. For example, a study of how a tropical bird budgets its time for feeding itself, feeding young, resting, preening, etc., might suggest a similar project with a related species — the observation techniques would be the same, and a comparison of how the tropical and temperate environments affect the use of time by closely related species could lead to some very interesting conclusions.

For information on migration throughout North America, breeding and wintering populations, and the Christmas bird counts sponsored by the National Audubon Society, look in *American Birds* (formerly *Audubon Field Notes*), published six times a year by the National Audubon Society in collaboration with the United States Fish and Wildlife Service. Bird organizations of the states and of even smaller regions publish reports on migration, distribution, and seasonal bird populations. All are interested in receiving records of unusual sightings and census results.

Bird-Banding, a quarterly, is published jointly by the banding organizations of this country; several regional banding magazines are published also. They all describe banding studies, and are excellent places to look for data on the speeds, routes, and distances birds travel on migration.

Three important North American quarterly journals cover all aspects of ornithology. Although most articles are very specialized, a look through any of these magazines will show you some of the major concerns of ornithology today, and may suggest topics you yourself would like to investigate. They are *The Auk,* published by the American Ornithologists' Union, known as the A.O.U., *The Wilson Bulletin,* of the Wilson Ornithological Society, and *The Condor,* of the Cooper Ornithological Society. At the end of every article in these journals you will find a list of cited literature that can help you locate more information on the topic.

Devoted strictly to the sport of bird watching is the bimonthly magazine *Birding,* published by the American Birding Association. It gives information on good birding localities and where to find particularly unusual species.

(An address for each of these magazines is given in the Appendix on page 291.)

As you can see, there are many reasons for studying birds and many ways to do it. The following chapters will provide you with information on how and why birds function as they do. With this knowledge your enjoyment of seeing birds will grow and you will be equipped to make discoveries and draw your own conclusions about the birds around you and their role in the environment.

Chapter Three

An artist's conception of HESPERORNIS, *a bird that lived 120 million years ago. Note the toothed bill, absence of wings, and webbed feet.*

Origin, Evolution, and Speciation

How the birds we know today came into being is a fascinating story, and we can better understand their appearance and behavior by knowing how they have evolved and adapted to their present roles. Knowing the past also helps us determine relationships between modern families and species, so that we can understand which aspects of behavior, structure, and appearance were developed by one species to meet the challenges of its particular environment and which aspects are shared by other members of a family.

Fossil History

Fossil evidence indicates that birds are descended from reptiles, with which they still share many characteristics. Birds, of course, have feathers and are warm-blooded, unlike all modern reptiles, but the two groups have many similarities of bone, muscle and joint structure, blood cells, and egg type. Precisely how birds developed from reptiles remains very much a matter of guesswork, in part because we have so few fossils of the earliest birds. Bird skeletons are fragile, and few birds happened to die in places where their bodies would be covered by the layers of mud or silt necessary to preserve the skeleton or its impression. As you might expect, most of the oldest bird fossils are of water birds that had strong bones and lived in environments where their skeleton had the greatest chance of being preserved.

Several types of now extinct reptiles could fly. Some had wings made of flaps of skin held stretched by the forearm (similar to bats); other lizardlike animals, such as the one shown overleaf, had flattened ribs covered by skin, enabling them to glide, but not flap, like a flying squirrel. However, there is evidence that none of these flying

An ancient reptile with expanded, flattened ribs that enabled it to make gliding leaps.

reptiles were ancestors of birds: they lacked the clavicle (wishbone) found in all birds, and they lived at a time when some birds already existed.

The earliest known bird has been named *Archaeopteryx* ("ancient wing"). Its fossils, found in German deposits believed to be approximately 155 million years old, indicate birds like the drawing opposite, with feathers except on the head and neck, short rounded wings with claws, feathers growing down the sides of a long, lizardlike tail, and teeth in the jaws. The skeleton resembles that of the small dinosaur, on page 24, that ran on its hind legs and used its forelegs to grasp prey. *Archaeopteryx* may also have lived on the ground, using its wings mainly to help catch prey rather than for flight, or in trees, using the claws in the wings to help it climb, and gliding from one tree to another. In either case, *Archaeopteryx* was not a very advanced flying machine — it did not have the bones or muscles essential to powerful flight and its bones were not hollow for lighter weight, like those of most later birds.

Feathers are extended reptile scales, but the reasons for their evolution into feathers are open to question. They may have been developed by ancestors of *Archaeopteryx* that gradually did more gliding and less jumping from branch to branch, or flapped their forearms to increase speed while running. Other theories are that some dinosaurs were warm-blooded and originally developed extended scales to keep warm, or that extended, moveable scales developed to keep cold-blooded dinosaurs cool, by shading the body from the sun's rays. In any case, use of these extended scales, or feathers, for flight only came later. Discovery of other early bird fossils will tell us more.

The next oldest known fossils come from the shale beds of western Kansas, once an inland sea. They are all 35 million years more recent than *Archaeopteryx*. One is *Hesperornis* ("western bird"), a toothed, loonlike bird five feet long and highly specialized for swimming. Its wings were reduced to two small bones. Another is *Ichthyornis* ("fish-bird"), shown on page 25, the first known bird with wings developed for powerful flight. It was a gull's size and presumably ate fish; whether it had teeth or not is undecided. Neither of these birds has any modern descendants, nor do most of the approximately 33 fossil species dating from this period. (A few specimens resemble modern loons, grebes, and rails, but are not definitely ancestors). The diversity of bird types already present 120 million years ago is impressive.

From about 60 million years ago we have fossils of birds from

An artist's conception based on fossils of ARCHAEOPTERYX. *Note the toothed jaw, clawed wings, and long bony tail, unlike the boneless tail of modern birds.*

A possible ancestor of birds, this reptile ran on its hind legs; the forelimbs were free to grasp prey or to flap to increase speed while running.

several modern families, including the herons, ducks, vultures, hawks, grouse, cranes, rails, sandpipers, and owls. None of these species still exist, however. Other fossils represent types that have disappeared, such as *Diatryma*, a flightless, heavy-bodied, seven-foot bird; its fossils have been found in Wyoming, New Mexico, New Jersey, and Europe. It may be distantly related to the rails.

Between 28 and 12 million years ago nearly all the families of larger North American birds developed, and by 10 million years ago there were even some songbirds. These, the passerines, are considered the most recently evolved birds.

One million years ago many species alive today existed, mixed with others that were shortly to become extinct. Since then, the earth has experienced several widespread changes in climate, especially periods popularly known as "ice ages" when large areas became colder and drier. These drastically altered many environments, causing the extinction of many birds that had lived in them, and created opportunities for evolution and expansion of the newer bird families, which took over some of the roles left vacant by the older, highly specialized birds that could not adjust to the changes. By the end of the most recent period of climatic change, about eight thousand years ago, all the birds that exist today had probably evolved.

Evolution

The evolution of new species, or of entirely new types of animals like birds and mammals from reptiles, is very slow and is difficult to visualize. The process is easier to demonstrate when it can be seen operating on a smaller scale. Charles Darwin (1809–82) was the first person to fully describe the process of evolution, in his *The Origin of Species* (1859). When studying the birds of the Galapagos Islands he found perfect examples for the ideas he later formulated.

The Galapagos Islands, on the Equator several hundred miles west of South America, are hot, rocky, and dry, seemingly inhospitable to any life. Reachable only by sea or air, they are populated with the descendants of plants and animals that can swim, fly, or drift through either medium to accidentally reach them; most of the plants and animals that came thus upon these islands perished, but some survived. For creatures that could find food, the Galapagos were a sort of paradise — they found very few or no natural enemies and they could multiply to the limits of their food supply. What Darwin found was a strange assortment of animals unlike any on the South American mainland. There were swimming lizards, flightless birds, and other birds with highly individual ways of getting food. All had ancestors on the mainland, but no surviving similar forms there.

Most remarkable on the Galapagos were the group now called Darwin's finches. Descended from some songbird blown off track, there

An artist's conception of ICHTHYORNIS, *the earliest known strong-flying bird.*

Darwin's finches have bills ranging from massive seed-crushers to thin insect-pickers.

are now fourteen distinct species, and they are noteworthy because each one has acquired a different way of feeding and a different type of beak best suited to it. Several are shown opposite. Each species specializes in feeding either on seeds, insects, buds, or nectar. One has made an even greater specialization — it has learned to probe the bark of trees for insects by carrying a twig, its own bill being unsuitable for that operation (illustrated on page 81). Darwin's finches are a perfect example of evolution — the gradual adaptation over many generations to utilize a niche in the environment. The development of even small evolutionary changes may take hundreds of thousands of years (the reduction from four toes to one in the foot of the horse took approximately 48 million years). We don't know how long the finches have been present on the Galapagos.

How does evolution work? Evolution means change, and in living creatures change comes through reproduction. In every case of sexual reproduction two sets of chromosomes containing many genes join to form a new set in the offspring. The genes are in charge of reproducing every feature of the parent, but when two sets come together in reproduction sometimes differences — mutations — are created. Mutations occur frequently in nature, but most are so minor or inconspicuous that we never notice them. In birds they may involve small changes in color or patterns, their internal structure, or things affecting behavior. Most of these mutations probably have no effect on the life of the mutant individual. Sometimes the mutation makes life more difficult; in these cases the mutant does not usually survive to pass the characteristic on to the next generation. If the mutation is something that gives it an advantage over other members of the species, then it is likely to be passed on to the next generation. The important thing is that these mutations are completely random: there is no preconceived design or natural course that they take.

What makes the evolutionary history of any species seem so logical and preconceived is the other important factor in evolution — natural selection. In every case the individuals that are going to survive and reproduce will be those best adapted to their niche, that is, those best able to find food, breed, and avoid enemies. If a particular mutation is going to give a creature any advantage over its fellows, natural selection pressures will work in its favor and this mutation will spread in succeeding generations. If the mutation is significant enough, these descendants will become a species distinct from their ancestors. In some cases the two species will coexist, each exploiting a slightly different niche, or in different localities. (A species, in fact, is defined as the population or set of populations whose members will breed only with one another.) This was the case with Darwin's finches, where random mutations created a variety of beak types, each useful for taking a particular food.

In studying the effects of evolution on closely related forms, we always look for the adaptive significance of the features setting them apart. The adaptive significance of the various bill shapes of Darwin's finches is that they enable each species to take a certain kind of food most efficiently. Nearly every feature of a bird's body and behavior has some adaptive significance that helps it exploit its environment more successfully. When looking at birds, consider the benefits each feature gives a bird, and why natural selection allowed it to evolve. Later chapters will consider each topic of bird biology from this adaptive point of view.

Darwin's finches also illustrate what is called "mosaic evolution," where some aspects evolve faster than others. The plumages of the finches are very similar, while the bills are distinct. There was little advantage in the evolution of different plumages, so they remained stable while the bills — in comparison — were evolving rapidly.

Availability of unexploited niches is one of the basic requirements for evolutionary progress, and has allowed the growth of whole new orders of life. If there had already been a wide range of small land birds on the Galapagos, there would have been no room for the Darwin's finches to expand into different species. When several forms evolve to occupy smaller segments of a wide ecological niche, as the Darwin's finches have done, it is known as radial adaptation.

The only way new species of birds can evolve is if the population containing a mutant gene is isolated from other populations lacking the mutation. Unless isolated, this small mutant population would quickly be "swamped" and the mutation would disappear. The evolution of distinct species requires very long periods of time. For this reason, a group of islands, like the Galapagos or Hawaiian islands, is an ideal natural laboratory for evolution. The small land birds rarely travel, and slightly different forms, long separated from their common ancestor, may be found on each island of the group.

Even birds that are now widespread on a mainland or several continents must have been at some time in the past isolated from near relatives. Changes in climate or land features may create "islands" separating different populations of a species, allowing them to develop in different directions. A look through any bird guide will reveal cases of recently evolved species that are still extremely similar, but now breed in different localities or different habitats. Among the warblers, the Black-throated Green, Golden-cheeked, Hermit, Townsend's, and Black-throated Gray are all "new" species, descended from one widespread species whose range was recently broken up, leaving isolated populations to evolve separately. The Yellow-throated and Grace's Warblers must have a similar history.

Some time after the period of isolation, separated populations may come together again. If they are sufficiently different in appearance, behavior, or ecology not to interbreed, they qualify as distinct species, that is, a population or set of populations whose members will breed only with one another under natural conditions. (Many distinct species will interbreed in captivity where normal barriers of geography, behavior, or ecology are eliminated.) If slightly different looking populations will interbreed where their ranges overlap, they are considered "subspecies." The Myrtle and Audubon's Warblers, formerly thought to be separate species, are now known to interbreed regularly where their ranges meet, and have recently been put into one species called the Yellow-rumped Warbler. Other widespread birds that have many slightly separated populations have evolved many subspecies, or races. The Song Sparrow, for example, has at least 36 subspecies, all so similar that no one now believes any should be treated as distinct species. Approximately 75 percent of all bird species have races or subspecies. If separated long enough, some may eventually become full species.

A few birds, the Screech Owl among them, come in two distinct colors that occur in the same area and interbreed regularly. These are called "color phases." There is usually no geographical isolation as in subspecies. The Blue and Snow Goose have recently been recognized as color phases of the same species. When color phases interbreed, their offspring look exactly like one or the other parent; but when subspecies interbreed, the offspring have a combination of features from each parent; Common Flickers whose parents were the subspecies "Red-shafted" and "Yellow-shafted," for example, may have either orange feather shafts or some reds and some yellows.

While evolution tends to create many different forms on the specific and subspecific levels, on the other end of the scale it sometimes leads to extremely similar but unrelated forms from different families. This is called convergent evolution and it shows the strong effect of natural selection on evolutionary history. For example, the Old and New World vultures, two unrelated families, both feed on the bodies of the dead animals they locate at great distances while soaring at high altitudes. The physical requirements for such a mode of existence are the same in both New and Old Worlds, and natural selection has favored birds that could most effectively meet them; both these vulture families are therefore large birds with broad wings and tail, ideal for effortlessly riding thermals for several hours while searching for food, and both have unfeathered heads and necks, because the naked heads are easier to rub clean of food or parasites picked up as the birds feed. Only recently have the differences in the two groups of vultures been recognized; the Old World Vultures are part of the hawk and eagle family, the New World Vultures are believed by some to be more closely related to the storks.

Books showing birds from different parts of the world will give you many examples of convergent evolution. Compare the North American Tennessee Warbler, of the New World wood warbler family, with the European Willow Warbler and Chiffchaff, both from the unrelated Old World warbler family, or the now extinct Great Auk of the North Atlantic with the penguins of the southern hemisphere, shown opposite. In each case the converging birds inhabit similar environments where they are subject to similar natural selection pressures and have evolved to the one size, shape, and color pattern that is most successful in that ecological niche.

Taxonomy and Systematics

Scientists have arranged all living things in a system that indicates, as far as known, how they are related and the order in which they evolved. Creating this system, or placing an animal or plant in it, involves two closely related and overlapping studies. Taxonomy is the identification and naming of an organism. Systematics, or classification, is the determination of how it relates to other organisms. The taxonomy and systematics of birds have been worked out in greater detail than those of any other group of animals, but this is still a lively aspect of ornithology. New research may give a better picture of the relationships of certain groups of birds and whether certain forms should be considered distinct species or only subspecies.

Convergent evolution: the Gentoo Penguin of the antarctic region and the now extinct Great Auk of the north Atlantic. The auk probably fed entirely on fish, while the Gentoo Penguin's diet is largely crustacean.

The first requirement of any classification system is that every known form have a name that will be understood by scientists working in any language. The need for a system naming all known forms of life has long been recognized. In the seventh century B.C. the Assyrians developed a bird classification system based on habitat. Aristotle, as we have noted, described over 170 bird species three centuries later. Pliny the Elder, a Roman naturalist of the first century A.D., classified birds by foot structure. Fifteen hundred years later, Europeans started arranging birds based on apparent similarities.

The system used today was developed by Linneaus in 1758. He gave every known species a name made of two words, usually derived from Latin or Greek. The first word is the name of the genus, a group of closely related species; the second name refers to the particular species. No other creature in the animal world may share that combination of names. Since the development of the subspecies concept in the nineteenth century, some animals have been given a third name, after the species name, indicating the subspecies. The name of the genus is capitalized, the name of the species and subspecies are not. Thus in 1766 Linnaeus named the already well known American Robin *Turdus migratorius.* All populations resembling the individuals from eastern North America first described to science belong to the subspecies *Turdus migratorius migratorius.* Individuals of other races, like the form that breeds in northeastern Canada, *Turdus migratorius nigrideus,* recognized in 1939, have different subspecies names. (When a species or subspecies is described from a particular specimen, that individual is called the "type specimen" and is carefully preserved in a museum.)

Each genus is made up of the species considered most closely related; all members of a genus are descended from a shared ancestral species. Some birds have no very close relatives and are in a genus by themselves. The current trend in ornithology has been to merge genera that seem closely related. McCown's Longspur, for example, formerly *Rhynchophanes mccownii,* a genus of one species, has recently been put into the same genus as the other longspurs and is now *Calcarius mccownii.*

All the genera sharing certain characteristics belong to one family; ornithologists currently recognize about 170 families. The family is the most useful category after the species: its borders are usually clearer and less subject to change than the genus. Today birds are rarely moved from one family to another, nor are families often merged. Characteristics used in assigning birds to a family include arrangement of wing feathers, patterns of scales on the legs and feet, bill shape, and some internal structures. Recent taxonomic research has tried to use all available information, including behavior, nest, eggs, and development of young. The composition of egg-white proteins, eye-lens proteins, and blood plasma has also been used, based on the assumption that these are "conservative" characteristics not likely to change, while behavioral or structural characteristics, which evolve more rapidly, do not show underlying relationships as clearly.

The scientific names of families end in *"-idae,"* like Anatidae, for swans, geese, and ducks, or Alcidae, for puffins, murres, and auks. These names are often converted to terms like "anatid" or "alcid," which are handier and more specific nouns than "swans, geese, and ducks" or "waterfowl." Within each family, the genera and species are arranged according to closeness of relationship and to their supposed order of evolution, based on fossil, anatomical, and distributional evidence.

There are larger groups, called orders, that consist of all the families more closely related to one another than to any other group of birds. Families of each order are presumed to have a common ancestor. Ordinal names of birds end in "-iformes," such as Anseriformes, containing the Anatidae and Anhimidae, a small South American family called Screamers, or Falconiformes, for the five families of birds of prey active by day, or Strigiformes, the two owl families, Tytonidae, the Barn and Grass Owls, and Strigidae, the typical owls.

The largest order is the Passeriformes, or perching birds. It includes more than 5000 of the world's 8600 to 9000 species, in 69 families. This collection of families is generally referred to as the "passerines," and includes all the families we think of as songbirds. One of the key characteristics they all share is a foot with three toes pointing forward and one pointing backward, adapted for perching. The passerines are the most recently evolved order of birds and seem to be continuing their evolution while earlier orders have stabilized.

All the orders and families of birds are arranged in a standard sequence reflecting their believed time of origin, with oldest groups first. With a few exceptions the sequence is followed by all ornithologists and all bird books. In some cases this order is based more on tradition than definite information. The penguins, for example, are listed first, but they are surely not the most "primitive" order of birds; their loss of flight, use of wings for swimming, and adaptations for survival in the challenging Antarctic environment are the result of a long evolution. However, no one has decided where else to put them, so penguins remain with the other flightless orders, ostriches, rheas, kiwis, etc., although there is no suggestion of a close relationship with any of these groups. Similarly, the orders of loons and grebes are placed next to one another and superficially resemble one another, but they are no longer thought to be closely related.

Within the Passeriformes, two sequences are widely followed. Americans put the Fringillidae (cardinals, finches, buntings, and sparrows) at the end, because they are believed to be the most recently and rapidly evolved family. The fringillids include about 375 species, are found throughout the world except in the Australian region and some oceanic islands, and occur in nearly all habitats. (The family Fringillidae is thought by some to contain birds of two separate family origins that have come to resemble each other by convergence. Cardinals, buntings, sparrows, and towhees are sometimes put in a distinct family, Emberizidae, leaving the finches, siskins, and crossbills in Fringillidae.) Some European ornithologists place the Corvidae (crows, magpies, and jays) and the related families of bowerbirds and birds-of-paradise at the end, because the corvids are considered the most intelligent of birds and the bowerbirds and birds-of-paradise have extremely elaborate courtship feathers and displays.

The order is the broadest classification usually used in ornithology, but still larger groupings can be used to specify the place of any birds within the realm of all living things. Thus, *Turdus migratorius nigrideus*, the American Robin of northeast Canada, can be identified as follows:

Kingdom:	Animal
Phylum:	Chordata (all animals with either a notochord or a true backbone)
Class:	Aves (birds)
Order:	Passeriformes
Family:	Turdidae
Genus:	*Turdus*

Species:	*migratorius*
Subspecies:	*nigrideus*

The table of contents of any bird guide will show you the arrangement of orders and families used by American ornithologists. After using the guide a while you will learn the sequence easily.

Some of the most recent changes in species and subspecies have not yet appeared in the guides. This book uses the new names, which reflect the most up-to-date thinking on the true relationships of certain forms. These decisions are made by the Committee on Classification and Nomenclature of the American Ornithologists' Union. The committee includes the foremost experts on taxonomy and systematics. Decisions on "lumping" forms previously considered separate species but now known to interbreed, or "splitting" those now recognized not to interbreed, are made only after a very careful review of all information available. When two forms have been lumped, like the Myrtle and Audubon's Warbler to create the Yellow-rumped Warbler, the old names can still be used, if you clearly mean only that subspecies. Following are some name changes that may not be found in your field guide.

New Name(s)	Old Name(s)
Great Blue Heron	Great White Heron, now considered a color phase of the Great Blue Heron
Brant	Black Brant, now a subspecies
Snow Goose	Blue Goose and Snow Goose, considered color phases of the same species
Green-winged Teal	Teal (the Eurasian form) and Green-winged Teal (the North American form), subspecies of one species found throughout the northern hemisphere
Red-tailed Hawk	Harlan's Hawk, now a subspecies
Thayer's Gull	formerly considered a subspecies of the Herring Gull
Common Flicker	Yellow-shafted Flicker, Red-shafted Flicker, and Gilded Flicker, all interbreed
Willow Flycatcher and Alder Flycatcher	Traill's Flycatcher, divided into two species

Tufted Titmouse	Black-crested Titmouse, now a subspecies
Bushtit	Common Bushtit and Black-eared Bushtit
Yellow-rumped Warbler	Myrtle Warbler and Audubon's Warbler
Northern Oriole	Baltimore Oriole and Bullock's Oriole
Boat-tailed Grackle and Great-tailed Grackle	Boat-tailed Grackle. The southwestern form is now considered distinct
Savannah Sparrow	Ipswich Sparrow, the form which nests only on Sable Island, Nova Scotia is considered a subspecies
Seaside Sparrow	Dusky Seaside Sparrow and Cape Sable Sparrow, like the Ipswich Sparrow, are considered isolated races of the widespread species
Dark-eyed Junco	Slate-colored, White-winged, and Oregon Junco.

The English names of several other species were recently altered for the sake of clarity — to prevent confusion with other Western Hemisphere birds, to bring them into conformity with international usage, or to avoid misleading taxonomic implications. These changes do not reflect any systematic changes.

New Name	Old Name
Northern Fulmar	Fulmar
Short-tailed Shearwater	Slender-billed Shearwater
Flesh-footed Shearwater	Pale-footed Shearwater
Wilson's Storm Petrel, etc. ("Storm" is added to the names of all Hydrobatidae.)	Wilson's Petrel
Great Egret	Common Egret
Wood Stork	Wood Ibis
Fulvous Whistling-Duck, etc.	Fulvous Tree-Duck, etc.
Northern Shoveler	Shoveler
Black Scoter	Common Scoter
Merlin	Pigeon Hawk

American Kestrel	Sparrow Hawk
Upland Sandpiper	Upland Plover
Red Knot	Knot
Gray Catbird	Catbird
Common Yellowthroat	Yellowthroat

Chapter Four

A male Ruffed Grouse "drumming." Air passing through the feathers as the grouse beats its wings produces a sound like rolling drums, by which the bird communicates possession of a territory.

Feathers and Flight

OF ALL the characteristics of birds, none is more distinctive than feathers. All birds (except the young of some species) have feathers, and no other animals do. Whether feathers first developed to keep birds warm, to keep them cool, or for flight, they now serve all these purposes and many more.

Functions of Feathers

Feather structures, colors, and patterns *sometimes* show the very particular pressures put on a species or a subspecies by natural selection, but in other cases, no one knows precisely what purpose, if any, a feather color or structure serves. No one knows what advantage it is to a Blue Jay to be blue, or if there is any adaptive significance to the darker blue of its close relative, the Steller's Jay. Perhaps for jays color "doesn't matter." The gene that controls feather color may also affect something of real adaptive value, so that the color remains because the other characteristic is essential. Even if we don't know why the jay is blue, we can guess that its crest and bright white markings serve important communicating functions.

Feathers come in several forms: smooth ones cover the body, fluffy ones underneath keep it warm, and various fine, hairlike ones have purposes not yet understood. The most basic function of feathers is to protect the body underneath — birds have very delicate skins (although the skin probably did not become so delicate until after it was covered by feathers). Birds also have very high body temperatures, most between 104° and 112° F, compared with man's 98.6°, and could not survive cold weather without a protective layer of feathers. Like human clothing, feathers keep a bird warm in two ways: by their own weight or thickness, and by trapping air underneath. Muscles connected to the feathers allow a bird to fluff itself up to entrap more air when it is cold. When birds are hot, they compress the feathers to eliminate pockets of air.

Feathers also protect birds from water, which runs off the overlapping layers; to increase waterproofing, most birds also apply oil to the feathers from a gland at the base of the tail. A few water birds, including cormorants and the anhinga, lack waterproofing and must spread their soaked wings out to dry after swimming.

Many kinds of animals have developed powers of flight, but feathers have allowed birds to evolve the greatest variety of flight techniques. As described in Chapter 6, muscles, bones, heart, brain, and lungs all evolved along with feathers to produce a more efficient flying machine, but it was probably the superior evolutionary potential of feathered wings and tail that made birds the dominant group of flying animals, rather than the mammals, insects, or now extinct orders of flying reptiles. Some of the advantages of feathers for flight are lightness, regular replacement when worn, lost, or damaged, and individual attachment to muscles for greater maneuverability. Birds can ride effortlessly for hours on the wind or thermals, travel speeds greater than 100 miles per hour, fly more than 48 hours without resting, hover and fly backward, use wings to "fly" under water, and cover thousands of miles in a season. No bird does all these things, but feathers make them all possible.

Feather colors also serve many functions, and since colors and patterns are more easily modified during the course of evolution than the structure of feathers or other body parts (most subspecies are told apart by differences in color), we suspect these modifications are due to local environmental conditions, or the demands of a particular role in the environment. The examples of convergent evolution noted in Chapter 3 are proof of this.

Feather colors and patterns serve two functions: protection and communication. For birds that are the prey of animals or other birds, one means of protection is to be difficult to see. Many birds therefore have plumages that blend in with their usual surroundings for camouflage. A look through a guide book will show how many birds — ducks, grouse, quail, rails, shorebirds, owls, nighthawks, wrens, meadowlarks, and sparrows — have streaky brown plumages to match their environment. Some more brightly colored birds have camouflaging plumages when they are young and least able to avoid enemies.

Among species dependent on camouflage for protection, natural selection quickly eliminates individuals less well camouflaged. This in part explains how races of certain widespread species evolve to fit their own area. Desert races of the Song Sparrow, for example, are lighter colored than those of humid areas — the light plumage blends in better with the plants and soil of dry regions. Some birds seem to "know" where they are camouflaged: certain African larks whose plumage matches only some of the soils in their environment will never land on a nonmatching soil, no matter how often they are flushed.

Some birds of prey that depend on not being seen by their prey as they approach it are also camouflaged; this camouflage also protects hawks and owls when they sit on their nests and do not want to be disturbed by the smaller birds that sometimes bother them.

Not all camouflaging plumages are brownish, of course. The black and white backs of woodpeckers often make them hard to find on a tree trunk. Ptarmigans, living in the far north or high on mountains, are brown in summer but white in winter to match the snow. Camouflaged birds of leafy environments like the vireos and warblers naturally are mainly greens and yellows. Often you can predict a bird's environment by looking at its plumage.

A bold, "disruptive" pattern can also be camouflaging. The stripes across the Killdeer chicks shown below, or on the Semipalmated Plover, for example, break up their normal outline when seen

Disruptive coloration: the bold black and white stripes on these Killdeer chicks break up their outline in a pebbly environment.

against a stony shore. Even the bright but splotchy plumage of birds like the male Rose-breasted Grosbeak are very difficult to see when the bird is sitting still in a tree where light and shadow form a splotchy pattern in the leaves.

Many shorebirds have dark upper parts, where they receive the most light, and light underparts, where they get the most shadow. By reducing the effects of shadow they are harder to see on a beach than were their entire bodies one color. This is called "countershading."

Startlingly visible plumage patterns can also protect birds. Many dark or dull-colored birds show flashes of white on or near the tail when they take off. This has evolved separately in many orders and families of birds. You can see it in shorebirds, doves, nighthawks, the Horned Lark, pipits, the Rufous-sided Towhee, juncos, and many others. The sudden flash of white may alert others nearby and also surprise the predator, directing its attention away from the bird's body; if the predator grabs on to the bright tail feathers, the bird may still escape. The white edges of the Blue Jay's tail, the white rump on flickers, and even the little white spots on a robin's tail may serve this purpose. A look through your guide book will show that these distracting flashes are most frequent and highly developed on birds that fly up suddenly from the ground.

You will have noticed that in some groups of camouflaged birds the sexes are alike, or nearly so. In others, like the ducks, males are very brightly colored, not at all camouflaged. Obviously, for male ducks bright plumage serves another purpose more important to survival than does camouflage. It has been found that the similarly plumaged females in the genus *Anas,* which includes the Mallard and the Pintail, cannot be told apart by the males, so selection of an appropriate mate depends on the female. Males have therefore evolved highly distinctive plumages and displays that make it easy for the females to recognize a male of their own species. Of course, even among birds that can recognize the opposite sex of their own species, the different plumages of the males may make it easier. This can be important for birds nesting in areas with a short favorable season — they can save time by seeing quickly who is an eligible mate without going through a time-consuming courtship.

Displaying plumage is an important part of courtship for many birds; attracting a female and driving away other males from the

Males of the Greater Bird of Paradise gather to display in bare trees in the New Guinea jungle. They bend forward, arch the back, raise their wings, and let the long flank plumes completely cover their body while they stand immobile or slightly quivering, as the females watch from the surrounding trees.

territory or nest site are usually closely related, and the same feathers and displays are often used in both activities. The response by the second bird to the display tells the male whether he should continue to be aggressive or begin the next stage of courtship. Some birds, like the egrets, have special feathers used only in courtship displays and dropped when courtship is over; the huge train of the peacock and the spectacular feathers of birds of paradise likewise serve only this purpose. Most birds, however, utilize their normal body, wing, and tail feathers in displays. Male Red-winged Blackbirds that had their red and yellow shoulder patches covered in an experiment were much less successful in defending a territory or gaining a mate. White-throated Sparrows of either sex that have the little patch of yellow between the bill and the eye are always able to drive away individuals lacking the yellow. The bright patterns of birds like male orioles and tanagers are easy to see when they fly from tree to tree; they may serve as a kind of advertising, to be interpreted differently by males looking for empty territories and females looking for a mate.

You can often see birds displaying. When courting, male pigeons and House Sparrows both fan their tails and drop their wings in front of females. In spring male Common Grackles show off their shiny plumage, and male Ruby-crowned Kinglets raise their crest when excited or chasing away other males. (The variety in bird displays is discussed in Chapter 8.)

Even with birds you don't see displaying, you can guess which parts of their plumage they use. In similar groups like the sparrows, the markings that are most distinctive, usually on the head, probably play a role. Even in the vividly colored warblers, the most distinctive markings are often around the head. Among the small, brown-streaked sandpipers, the only one to use its rump in displays is, predictably, the one with the most visible, contrasting rump — the White-rumped Sandpiper.

Besides specific displays, there are other uses for highly visible plumage. Consider how many seabirds are partly or almost entirely white, the color visible at the greatest distance in all light conditions. Seabirds with much white plumage include albatrosses, shearwaters, petrels, tropicbirds, gannets, gulls, and terns. They all feed offshore or at sea on fish, crustacea, or refuse which they must discover by searching a large area while in flight, since schools of fish and other food items move around and will not be found regularly in the same place. By being white, a feeding gull is most easily seen by other gulls far away; they can stop searching and join the one that has found food enough for many. Because they are visible as far away as possible, the birds can spread out farther and search a larger area with a greater chance of finding food. On some of the small, otherwise dark petrels, the white rump may serve this purpose. (Why, then, are some petrels and other seabirds *not* light? In some cases we don't know. Some of the all-dark petrels feed and search for food in groups, rather than spreading out individually, so seeing a feeding petrel far off is less necessary. The jaegers have little white, but they often depend on other light-colored seabirds to discover and catch fish, which the jaegers then steal. Cormorants, all dark above, locate their food while on the water or swimming under the surface; they would not be visible at a distance in any color.)

The colors of bird feathers may sometimes have a function separate from communication or protection from enemies. The black wing tips of so many large white birds (White Pelican, Gannet, Snow Goose, Wood Stork, White Ibis, Whooping Crane, many gulls and terns) contain a pigment which makes that part of the feather more wear resistant. You can see proof of this in gull wing feathers: if you find an old one on the beach you may see that its white outer tip is entirely worn away to the beginning of the black. The pinkish buff color of many desert birds may not have developed just for camouflage — it also helps insulate the bird's body better than any other color from strong desert light and from the extreme heat and cold of desert climates.

When you are close to a flying bird you can sometimes hear the sound of air rushing through its wing feathers. Some birds have evolved feathers that make louder noises and are an important part of display. The Ruffed Grouse has a "drumming" display in which it stands and rapidly beats its wings; the wings do not hit the body, but the air passing through them produces a sound like the roll of drums. The wings of the American Woodcock each include three narrowed flight feathers which produce a variety of whistling and twittering sounds that combine with calls on display flights. Hummingbirds, of course, are named for the sound their wings make in normal flight; some also have stiff outer tail feathers that produce sharp noises when suddenly spread at the bottom of a courtship dive.

Selective pressure has caused other birds to evolve in the opposite direction. Since owls often locate prey by very faint sounds that might otherwise be obscured by the rustling of their own feathers, and then must make a silent approach to the unsuspecting victim, their wings and bodies are heavily lined with soft feathers that muffle sound.

Feather Structure and Arrangement

To serve all these functions, several specialized types of feathers have evolved from the scales that covered the ancestors of birds. The drawing on page 46 gives the names of the external body parts of a bird and the names of the feathers that cover them. The structure of a feather has only a few parts, but these are modified on different parts of a bird and in different species to perform different functions.

Examine one of the large feathers, from a wing or tail, of any bird. It has a central shaft, strong but flexible, running its entire length. Where it is attached to the body the shaft has nothing growing out of it. It is hollow inside, making the feather lighter and better adapted for flying, when a bird wants as little extra weight as possible.

On each side of the shaft is a flat surface made of many interlocking rows of barbs. You can separate them and then, by pressing above and below the separation, join them together just as a bird does with its bill. The drawings opposite shows the feather parts and how the tiny parts of a feather work to form an even, unbroken surface.

The feathers which form the outer covering of a bird's body are called contour feathers. They include the feathers of wing and tail. All have the structure just described, but the feathers covering the actual body often have interlocking barbs only at the ends, on the part not covered by an overlapping feather; unconnected lower barbs leave more air space for insulation.

Some adult birds, especially those living in cold climates, have a layer of down feathers underneath the contour feathers. Down feathers, such as the one shown on page 48, have a very short shaft with many non-interlocking barbs sprouting in a tuft from the end of the shaft. This loose structure, although quite light, traps lots of air near the bird's body and is warmer than a heavy layer of contour feathers. Young birds are covered with down before they grow contour feathers. (Down is the most efficient protection against cold ever developed — man has not been able to invent a better material, and the

Major body regions of a typical passerine, the Lark Sparrow.

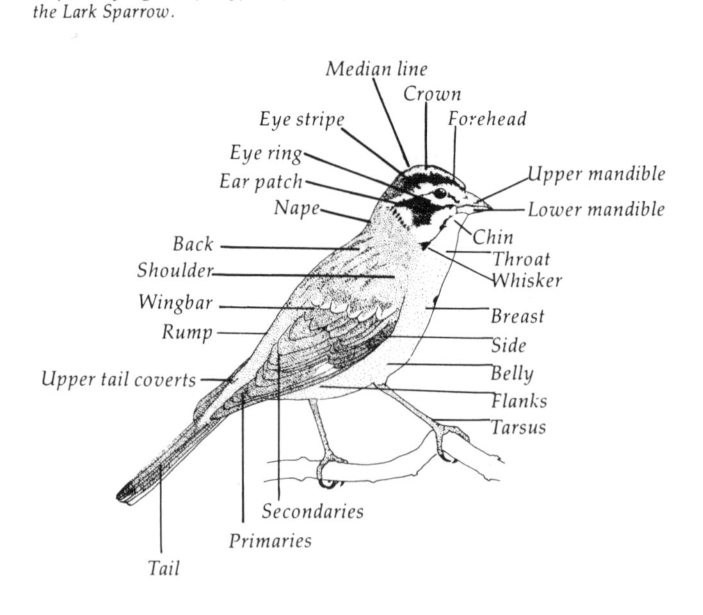

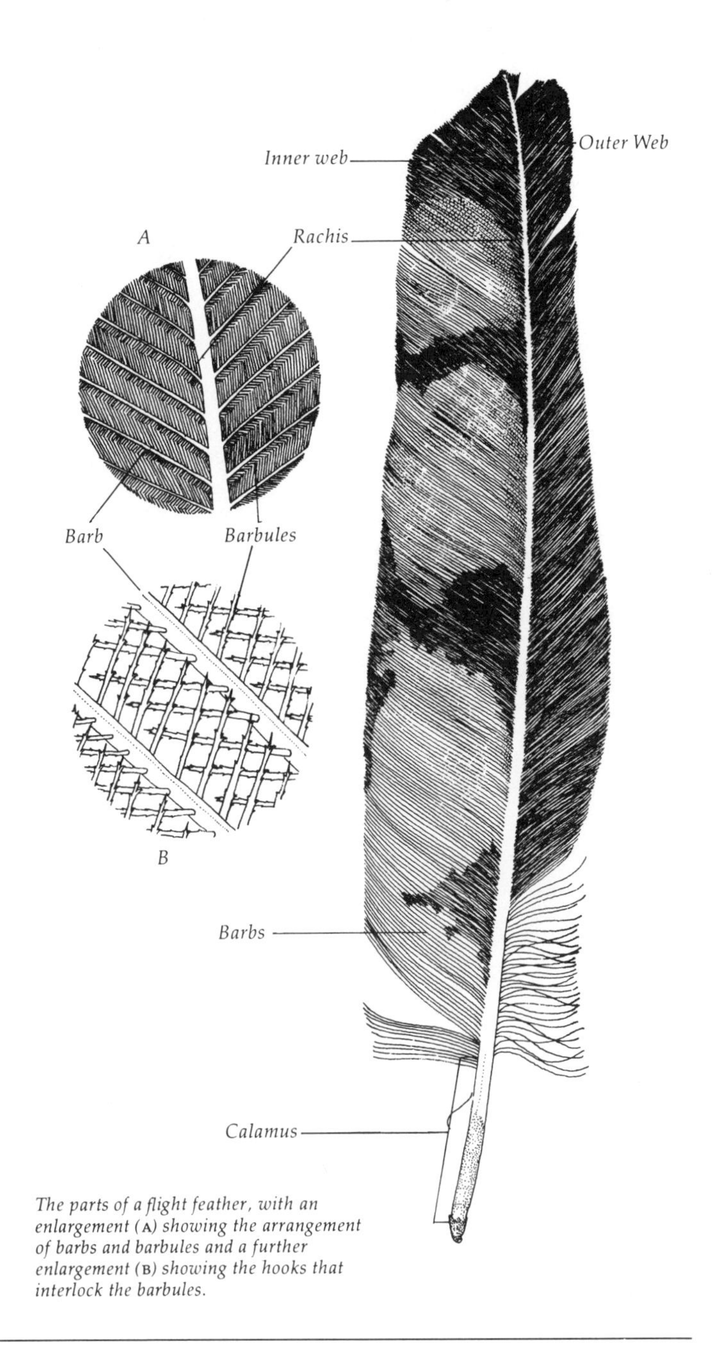

The parts of a flight feather, with an enlargement (A) showing the arrangement of barbs and barbules and a further enlargement (B) showing the hooks that interlock the barbules.

best cold-weather coats, blankets, and sleeping bags are filled with down, usually from ducks.)

There are other, more specialized types of feathers; some are shown below. "Powder down," probably derived from disintegrating down feathers, occurs as a light dusting on some birds. It is especially well developed in herons and bitterns, where it may aid in removing fish and eel slime from the plumage. Semiplumes help insulation and increase water birds' bouyancy. Filoplumes are small hairlike feathers with a few barbs at the tip of the shaft; they occur among the contour feathers, but their function is unknown. Some birds have feathers around the bill with only a few barbs at the base and the rest of the shaft naked. These are called bristles, and many birds that catch flying insects, like flycatchers, nighthawks, and the American Redstart, have them, but it has not been proved that they help in catching insects.

The feathers of wing and tail are naturally the most important ones for flight and, while keeping the same structure, have evolved the greatest variety of shapes, reflecting the needs of each species. There

Four specialized types of feathers: from left to right, filoplume, bristle, semiplume, down feather.

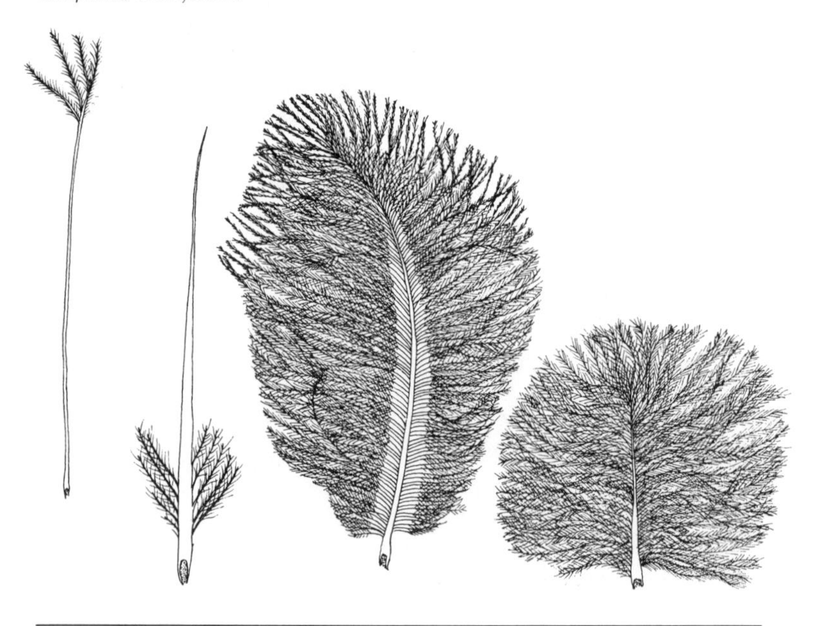

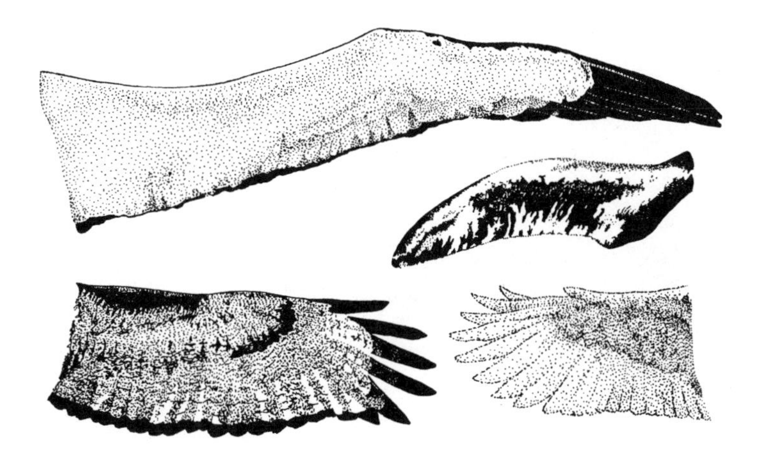

Wings of different shape (not drawn to scale): the five-foot-long, thin wing of the Wandering Albatross is adapted to catching gusts of wind; the broad, rounded wing of the Red-tailed Hawk has a large surface area to get the support of thermals; the Black-footed Penguin's wing is streamlined to serve as a flipper; and the short wing of the Willow Ptarmigan is useful only for brief but fast flights.

can be variety even within a species; when some races are migratory and others not, the migratory races have slightly longer wings. In general, birds that fly little or not at all have short, rounded wings, while those that fly a lot have larger wings; the wings of fast-flying birds are long and pointed. Tail feathers are also modified by needs. Because birds that ride on thermals, like vultures and hawks of the genus *Buteo*, need as much surface area as possible, they have broad wings and broad, rounded tails. Hawks of the genus *Accipiter*, which hunt while flying quickly through trees where they must avoid branches and make sudden turns, have shorter wings but long tails that help in steering. A few characteristic wing shapes are illustrated above.

The drawings on page 50 of a Blue Grosbeak's spread wing show the different wing feathers. The most important feathers for flight are the long ones at the tip of the wing, called primaries. All flying birds have between 9 and 12 primaries on each wing. The number is the same within most families and some orders, so the number of primaries has often been used as a taxonomic characteristic. The majority of birds have ten primaries, but some familiar passerine families like the wood warblers, blackbirds, tanagers, and sparrows have

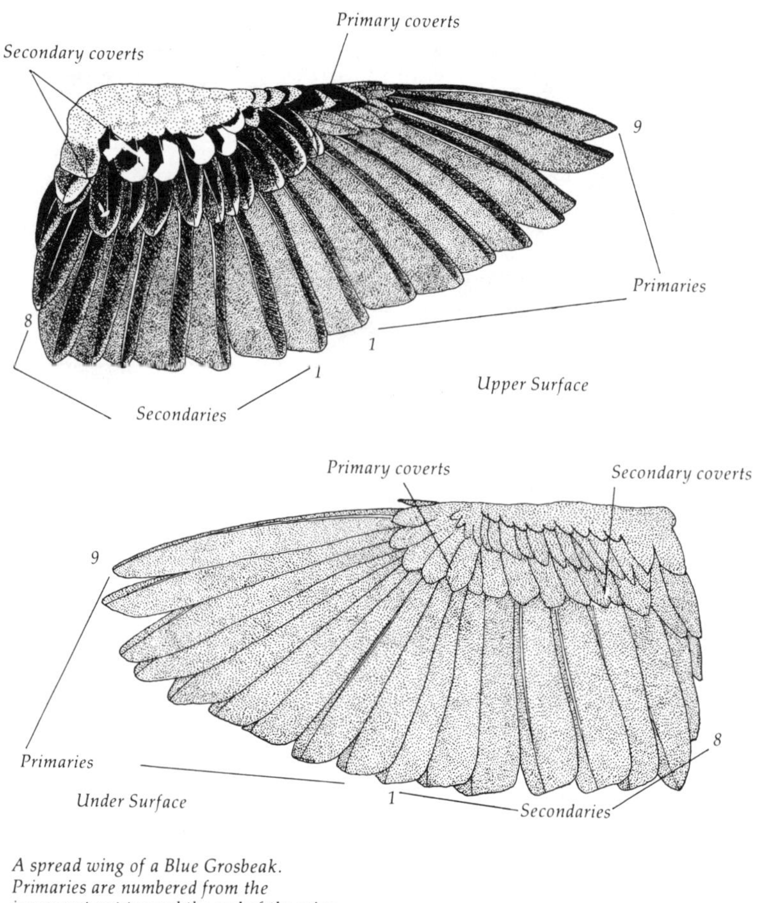

A spread wing of a Blue Grosbeak. Primaries are numbered from the innermost out toward the end of the wing; secondaries are numbered in the opposite direction. The rows of shorter feathers are primary and secondary wing coverts, called, from largest to smallest, greater, median, and lesser.

nine. Because these families have all evolved very recently, it is thought that newer families have reduced the number of primaries, rather than old families having evolved a higher number. Primaries are numbered from the innermost outward, because, it is assumed, families that have fewer primaries have lost them from the tip of the wing.

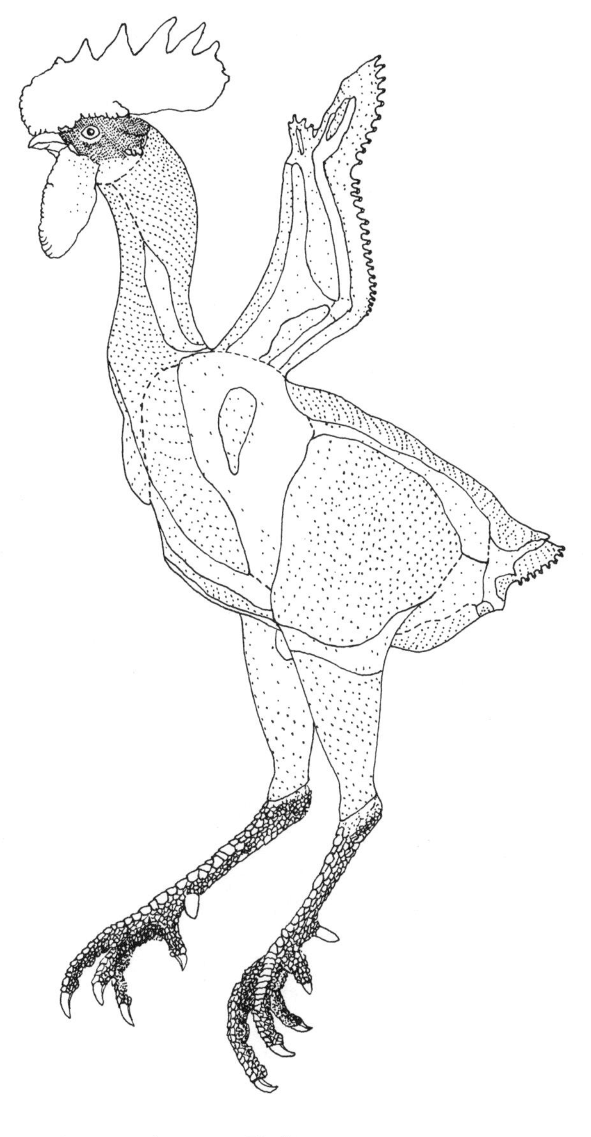

Feather tracts of a rooster. The lines drawn on the body define the areas of what are considered separate tracts (from Lucas and Stettenheim, AVIAN ANATOMY/INTEGUMENT*).*

The feathers that extend along the rear edge of the wing from the first (innermost) primary to the innermost part of the wing are called secondaries and are numbered from the outermost inward. The total number of secondaries depends on the length of the wing and is not as constant within groups as are the numbers of primaries. Hummingbirds, with very short wings, need only 6 or 7 secondaries to cover the space between the primaries and the base of each wing. Most passerines have 9 or 10. The number increases in the larger birds; the Andean Condor, with wings five feet long, has 25 secondaries, and the Wandering Albatross, with similarly long wings, has 32.

The tail feathers also vary in number and size. Their shape may reflect their function in flight, and they are often used in display. Some birds that climb tree trunks, like woodpeckers and creepers, have very stiff tail feathers that aid in support. Trunk-climbing birds that do not brace themselves with their tail, like the nuthatches and Black-and-white Warbler, have not evolved specialized feathers. Most birds have 12 tail feathers, in six pairs numbered from the center. Some birds have fewer than 12; grebes have 8 very short ones. Pheasants and other fowl-like birds often have more than 12. The famous display feathers of the male Peafowl ("peacock") are actually its upper tail coverts, the feathers on the back just above the tail, but it has 20 true tail feathers to help support this display; the female Peafowl has only 18 tail feathers.

The flight feathers are very important, but only form a tiny percentage of the total number of feathers covering even the smallest bird. The other contour feathers, down, filoplumes, and semiplumes are smaller and much more closely arranged. Small passerines average between 3500 and 5000 feathers, and even the tiny Ruby-throated Hummingbird has more than 1500. Water birds requiring heavy insulation have the greatest number of feathers. Nearly 12,000 were counted on a Mallard, while a Bald Eagle, several times larger, had only 7100.

Feathers are not distributed evenly all over the body. They grow in rows, or tracts, in very regular arrangements. You can see the arrangement on any plucked chicken like the one shown on page 51. Between each tract is a bare area completely covered by overlapping feathers. In a few groups, like the ducks, the spaces between tracts are covered with down. Only the ratites (ostriches and other large flightless birds) and the penguins have feathers distributed evenly all over the body, but this is believed to be a later evolution rather than the original arrangement for all birds.

Molt

After several months feathers may become worn, frayed, or lost. To avoid any dangers from a plumage in less than perfect condition, birds regularly replace their feathers in the process called molt. The feathers are dropped in a regular sequence at certain times of the year, and new ones grow in almost immediately. Molt may take a few weeks, or in the large birds that drop only a few feathers at a time, almost a year. Growing new feathers consumes a lot of energy and therefore usually takes place when a bird has no other important activities, like laying eggs, feeding young, or migration, to perform. Molt occurs at different times in different species, most often in the breeding area after young are independent, or following the fall migration. Experiments with a few species have shown that molt can be stimulated by changing the amount of daylight they receive. This

Feather tracts of the Scrub Jay. These views of the upper and lower parts show that feathers grow from only a small portion of the body surface (from F. A. Pitelka, CONDOR, *47, 1945.).*

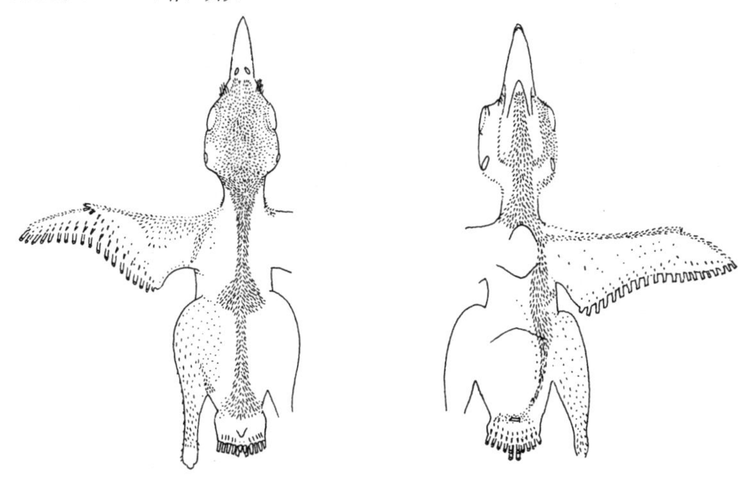

indicates that in natural situations birds may be stimulated to molt by the decreasing day length of fall and the increasing day length of spring; birds kept without changes in day length did not molt.

All birds are born with, or acquire shortly after hatching, a covering of down. As the young bird grows, the down is replaced by a set of contour feathers called the juvenal plumage, to which you can sometimes see bits of down still clinging. (In a few families, including woodpeckers, kingfishers, and some passerines, the young are born naked and go directly into this plumage.) During their first summer or fall, young birds molt all feathers except the primaries, secondaries, and tail feathers. In late winter or early spring some birds again molt all but the flight feathers. After the breeding season adult birds have a complete molt. To avoid the confusion caused by terms like "breeding" or "winter" for the plumages birds are often wearing in other seasons as well, the plumage acquired by the complete, post-breeding, molt is now called the "basic" plumage. The plumage some birds get from another, partial, molt is called the "alternate" plumage.

Some of the changes we see in plumage are not the result of molt, but of wear. The white feather tips acquired by the Starling in its prebasic molt are all worn off by spring, producing its shiny, all-dark plumage. Fading makes the plumage of some birds, like the Blue Jay, appear duller in spring than in fall and winter.

Some birds, especially large ones, do not breed in their first year and do not acquire the plumage of breeding adults until they are ready to breed. The Herring Gull goes through four prebasic molts before acquiring adult, breeding plumage. Bald Eagles often do not get white head and tail feathers until after their fifth year. Other birds breed in their first year without having acquired the typical adult plumage. This seems especially true of birds with red, orange, or black in their plumage — in spring you can often recognize pale first-year male American Redstarts, Northern Orioles, and Rose-breasted Grosbeaks.

At the end of summer you can tell that a lot of birds are molting. Some look ragged, missing wing or tail feathers; others, like the "confusing fall warblers," no longer have the bright patterns seen in spring, and you find more feathers lying about than in any other season. These are all birds that go through their prebasic molt before migrating. Some shorebirds that nest in the far north where the short warm season doesn't give them time to nest and then molt, wait until they reach their wintering ground. However, other shorebirds in the same environment nest and molt at the same time. The advantage of each system is not always clear. Even in closely related groups, like the almost identical looking *Empidonax* flycatchers, the Acadian and Hammond's Flycatchers molt on the breeding grounds while the others molt after migrating.

Some water birds like grebes, ducks, coots, and rails that can hide in marshes until they can fly again lose all their flight feathers at once. Most loons go through the flightless stage while on their wintering ground — the ocean or the Great Lakes — where they can escape danger by diving. In summer the males of many ducks including the Mallard have an alternate, "eclipse," plumage resembling the female's. They are flightless for part of this time, during which the camouflaging plumage may protect them. In late summer and early fall, male ducks are a splotchy mixture of browns and the familiar plumage.

The timing and energy requirements of molt have become popular research topics, often used to help explain how each species meets the challenges of its environment.

Feather Colors

The tremendous variety of colors found in bird feathers is the result of a unique set of evolutionary circumstances. There would have been no reason for the vivid colors and fine patterns many birds have, if birds could not see colors, but unlike mammals (except the primates) and reptiles, birds have color vision. The camouflaging patterns that conceal birds don't have to be perceived in color, but the bright colors that communicate things between birds would often be valueless if they couldn't be seen in color.

Feather colors are produced in two entirely different ways — by pigment and by structure. One of the commonest groups of pigments are the melanins, which produce dull yellows, reddish browns, browns, grays, and black. Other yellows, oranges, and bright reds are produced by carotenoids, another pigment group.

The other colors found on birds are the result of light reflected from the structure of the feather. The structures of feathers we see as white have no pigment and reflect all light. Colors in between white and black are produced by a combination of pigmented cells in the feathers overlaid with cells that reflect light of certain colors. All the blue colors on birds are the result of a layer of blue-reflecting cells over a layer of dark melanin-pigmented cells. Green feathers have the same structure, with a transparent layer of yellow-pigmented cells over the blue-producing cells, combining to create green. Dipping a blue or green feather in alcohol prevents the structural color cells from reflecting light, and reveals the color of the feather itself. A Blue Jay feather seen in the transmitted light of a microscope appears brown.

Iridescence, the changing glossy colors seen on grackles, Starlings, hummingbirds, and the necks of pigeons, is caused by twisted and flattened barbules on the feathers, which disperse and scatter light unequally so that its angle is constantly changing. An iridescent feather will take on different colors as you move it about in the light. A grackle or Starling in the shade, where it is not exposed to direct light, will lose its iridescence and simply appear black.

Most birds have colors that are both structural and pigmented, often on the same feather. Looking through a guide book will show how frequently and closely they are combined.

Occasionally bird feathers have more or less pigment than they are supposed to. When melanin is lacking, the feathers are white. This is called albinism, and is the most frequent of the pigment abnormalities; it may be partial, with only a few feathers or regions white, or complete. An excess of dark pigment is called melanism; still rarer cases are an excess of red or yellow.

Feather Maintenance

To keep the feathers in good condition and proper order requires regular care from the bird. All birds spend time preening, using the bill to reattach separated barbs and to smooth the feathers in place. You can most frequently see birds preening when they are not actively feeding. Some birds which seem to spend a lot of time just standing around, like gulls, are actually busy preening.

Bathing in water and dust also helps keep the feathers clean and in good condition. If you watch a birdbath or a shallow, slow moving stream, you will soon see how many birds depend on it. Late afternoon is often a favored bathing time. Notice the various ways different species bathe: some use their wings to splash water all over the body; others prefer to flutter in a thick patch of wet leaves. Hummingbirds, while in flight, press against a wet leaf or repeatedly dip into the water. After bathing, the bird will preen itself thoroughly. House Sparrows frequently bathe in dust, as will some other birds if water is not available. Dust can absorb excess oil or dirt in the feathers.

Many birds spend time sunning themselves. For birds like cormorants that have no natural waterproofing, this is essential; it may also benefit the feathers of other birds.

When birds preen, they rub their bill near the base of the tail and then over their feathers, spreading waterproofing oil from a gland located just in front of the tail. The gland is immediately below the surface; a few tiny feathers growing out of it serve as a wick. Without waterproofing, birds would become quickly chilled in rainstorms and water birds would be limited to coastal areas where they could get out of the water to dry, as cormorants must. Why cormorants, and a few other groups of birds that get their food from the water, never evolved an oil gland or other means of waterproofing is unknown.

Flight

Feather shapes are modified in several ways for different kinds of flight. The ideal shape of wings and tail for one kind of flight may not be useful for another. As we have seen, each hawk has evolved wing and tail shapes best suited to its hunting technique.

The first kind of flight developed by birds was gliding — coasting on wings held outstretched, going as far as height, wind, and speed of takeoff allow. Before birds developed the powerful muscles necessary for flapping, this was the only kind of flight possible. Today, birds that fly infrequently and only for short distances are primarily gliders. Pheasants and quail beat their wings rapidly to gain a little altitude when flushed, and then glide off on stiff wings. Their wings are short and rounded, and would not be very efficient for sustained flapping or for using winds and air currents. They are designed like the simplest paper airplane, which can go quickly for a short distance, but cannot catch the lift from breezes or thermals.

Many other birds use the gliding technique when they have already gained enough height and speed and want to save energy or descend. Swallows, swifts, and Starlings can all be seen gliding be-

These drawings, based on a series of photographs, show the extent of forward and backward motion in an eagle's wing flap.

tween periods of flapping, and ducks and shorebirds glide to lose altitude when landing.

Sustained flight requires much greater development of muscles and other body parts. Strong-flying birds show several evolutionary developments, including a streamlined body and wing shape to minimize air resistance; hollow bones and hollow feather shafts to reduce weight; internal organs arranged not to interfere with the development of large muscles used by the wing; and lungs, heart, and circulatory system that provide the rapid supply of energy required. (Some of these anatomical adaptations are described in Chapter 6.)

Flapping flight involves an up and down motion of the wings to keep the bird aloft and also a practically invisible forward and backward motion to make it move forward. The forward and backward aspect is shown clearly in the series of silhouettes of an eagle's wing flap above. A tern flying into the wind also shows the forward motion of the wings, which is something like rowing a boat. When flapping, or rowing a boat, the downstroke starts from its highest point by moving forward, then down and back, forcefully pushing the air or water. The upstroke moves forward with as little resistance to the air as possible.

Several mechanisms make a bird's wing more efficient at this than a boat oar. On the downstroke, the concave underside of the wing (like a scooped out paddle) grips more air, the flight feathers are held together so that no air escapes, and the wing is extended as fully as possible in a comparatively slow, powerful flap. Each outer primary provides forward propulsion like the blades of a propeller. On the upflap, the wing is partially folded and brought up quickly, and the primaries are more widely separated to create slots of air between them, lowering air resistance and turbulence. In many large soaring birds, including most hawks and eagles, the slots are especially wide because the primaries, broad at their base, become much narrower at the tip.

When flying into the wind, a bird must make adjustments. If it lowers the angle of the wing while facing into the wind, resistance is reduced but lift is also lost. If it raises the angle of the wing, it gains lift like a kite facing the wind, but loses speed.

Soaring flight uses wind currents and rising warm air to provide lift that would otherwise require the bird's energy. A soaring bird must have large wings to catch the lifting force of wind or thermals. No small birds can soar, just as a paper kite with the dimensions of a robin would be practically useless.

Soaring birds include hawks, eagles, and vultures, which all have broad wings and broad tails. They search for food on the ground while high in the air and must be able to spend hours aloft without exhausting themselves. Most of these soaring birds cannot fly for long periods by flapping, but travel far by riding a thermal to a great height and gliding to another thermal, as shown on page 60. Thermals are not produced over bodies of water or at night, so these birds are restricted as to when and where they can travel.

Seabirds have developed a different kind of soaring flight which

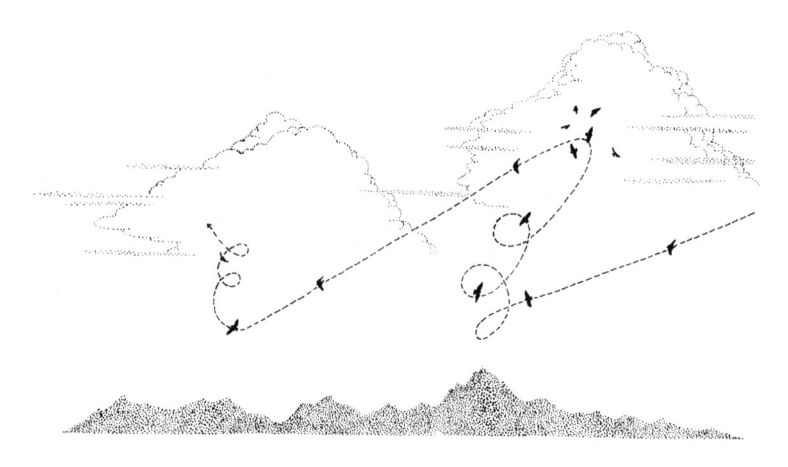

Many soaring birds, like this group of migrating Broad-winged Hawks, travel long distances without tiring themselves by rising in circles on a thermal and then gliding until they encounter another.

depends on wind. Albatrosses, shearwaters, and frigatebirds all have extremely long, narrow wings. Their kind of flight requires continuous winds, which do not carry them very high, but give them enough lift over the waves to make a long wind-pushed glide until they descend to wave level, when they turn into the wind and are again lifted like a kite for the next glide.

Many waterbirds, both soarers and flappers, depend on wind to become airborne, and often have difficulty taking off from the ground when they cannot drop off a high perch. Ducks like eiders and scoters, usually found on open water, as well as loons, grebes, and coots, must run on the water into the wind, flapping their wings, to take to the air. Other ducks, such as the Mallard, that live in marshes or small ponds where they can be surprised suddenly without room for a running takeoff, have evolved the ability to jump right into the air.

A few birds can hover in one place while flying. The most outstanding example, the hummingbirds, often seem suspended in air as they beat their wings 50 to 60 times per second while hovering in front of a flower. Hovering is usually done with the body in a vertical position, and uses muscles developed especially for this type of flight. American Kestrels can often be seen hovering over a field, preparing to drop on prey. Kinglets and a few warblers sometimes hover in front of a leaf to reach an insect, but they do not have the muscles necessary to hover for more than a few seconds.

Strong flying birds often travel hundreds of miles on migration without stopping, but rarely go at top speed. Passerines normally fly about 15 to 25 miles per hour, but can go more quickly for short distances when chased. Most ducks and geese travel between 40 and 60 miles per hour. Shorebirds go about the same speeds, but have occasionally been timed from airplanes at 110 miles per hour. The fastest birds are swifts and falcons. The large White-collared Swift of Central and South America has been timed at 150 miles per hour, and falcons like the Peregrine reach their greatest speeds when diving — "stooping" — on prey; speeds of 180 miles per hour have been recorded.

A few large birds like geese, cormorants, and pelicans are frequently seen flying in "V" formations or straight lines. The advantage of flying this way is still being argued, but each bird behind the leader may get a little bit of lift from the flap of the bird in front of it, or the flap of a bird in front may create a vacuum making the upstroke easier for the next bird.

Many birds have distinctive ways of flying. Woodpeckers are easily recognized by their typical roller coaster flight. Goldfinches also have a roller coaster flight, but its adaptive significance is unknown — the closely related siskins fly in a straight line. Ducks have a fast and straight flight; with their short tails they have difficulty making sharp turns. Many other birds can be identified by the way they fly. After watching them a while you can correlate the way birds fly with how they find food, where they live, and other basic aspects of their lives.

Chapter Five

The Everglade, or Snail, Kite has a curved bill adapted to extracting snails of the genus POMACEA *from their shells. These snails are the kite's sole food; it is a rare bird in south Florida, but widespread in Central and South America, nesting in loose colonies wherever the snails occur.*

Food, Feeding Habits, and Digestion

ONE REASON for the success of birds as a group has been their ability to feed on such a wide variety of plants and animals. Their universal distribution indicates that almost every environment contains food for some type of bird, and the variety in sizes and shapes of birds reflects their evolution in many directions to exploit different food resources.

Birds burn up energy very quickly, and the smaller they are the faster they burn it, so they must feed often to refuel. Land birds weighing 100 to 1000 grams (approximately 0.25 to 2.5 lb.) eat material equalling 5 to 9 percent of their own body weight every day, and with birds weighing 10 to 90 grams the figure may go from 10 to 30 percent. Hummingbirds may daily eat twice their weight in nectar. Some birds seem to do little except look for food all day, so feeding is the most easily observed aspect of bird behavior. Every species has a characteristic way of locating and taking food; some are unique and you will recognize them quickly.

When watching birds you will see that there are two parts to feeding. The first is locating the food item, whether it is animal or vegetable; the second is catching or picking it. For different birds and different kinds of food, each of these parts takes a different amount of time. A Red-tailed Hawk, for example, may spend many hours soaring over fields and meadows looking for a mouse, but can swoop down and catch it in less than a minute. A loon diving for fish may spot one quickly, but then have a long underwater chase to capture it. Plant-eating birds have the same situations — fruit eaters may have a long search for ripe fruit which is then speedily consumed, and jays may find acorns quickly but spend a longer time breaking the outer shell to get at the meat.

The differences in search and capture times lead to some general principles of feeding behavior that you can observe in the field. When search time is long and pursuit and capture time short, feeders are generalists, taking many kinds of food. A Red-tailed Hawk, for example, could not spend hours looking for a small rodent like a field mouse or chipmunk and then reject it because it was not a certain size or species. However, birds whose food is abundant and easy to find can be more specialized; locating food is not time consuming for them, so they can afford to be selective. Sometimes both principles work on the same species in different environments — Great Blue Herons in northern New York, where fish are not abundant, are generalists, taking any fish they can catch, while those feeding in the more abundant waters of Florida are specialists, taking fish only of certain sizes and species. Generally, when search time is short and pursuit is long (as for the loons) birds specialize in prey with the greatest value per time spent in pursuit — it isn't worthwhile for a loon to chase tiny fish; the chase expends more energy than a small fish supplies. Size makes a difference, too: for most small birds, food items are common and search time is short, so they are specialized feeders; larger birds with longer search times are usually generalists.

There are few plants or animals that do not provide food for some bird. From plants, birds take the seeds, fruits, leaves, buds, sap, and nectar. In the animal kingdom, birds eat insects and other arthropods, crustacea, the combinations of tiny marine organisms near the surface of the sea called "plankton," fish, amphibians, reptiles, other birds, and mammals. Animals too large for a bird to kill, up to the size of whales and elephants, are eaten by scavengers when dead. Other birds regularly eat bee's wax, and the Ivory Gull of the Arctic feeds on the dung of whales, seals, and wolves. Birds can be divided ecologically (but not systematically) by what they eat.

Insect Eaters

Birds of almost every order and family eat insects, spiders, and other land arthropods. Nearly all passerines are insect specialists. Many birds that feed mainly on plant material give their young insects, which contain more proteins and other nutrients essential to growth than do seeds or fruit. Sometimes birds are of considerable value to man in reducing crop-damaging insects or other pests.

Insects are taken in all their growth stages — as eggs, larvae, pupae, and adults — and from every habitat. Some are picked off the leaves and twigs by foliage gleaners like warblers, vireos, and orioles, others from the trunk and heavier limbs of trees by creepers, nuthatches, and the Black-and-white Warbler. Woodpeckers bore underneath the bark to get at insects living below the surface. Flycatchers watch for flying insects from a perch; they pursue individuals and return to their perch. Swallows and swifts locate insects while in flight; at night they are replaced by caprimulgids (nighthawks, the Whip-poor-will, and others) which, with bats, fill an otherwise vacant niche, exploiting night-flying insects. (There is little direct competition between caprimulgids and bats, since bats usually feed on insects smaller than the caprimulgids take.) Insects from the forest floor and in meadows and prairies are taken by many birds using different techniques. Meadowlarks take grasshoppers while walking on the ground; the American Kestrel pounces on them from the air. Water and marsh insects are consumed by birds ranging from ducks and rails to cranes. Shorebirds pick insects from beaches and mud flats.

Within closely related groups you may observe slightly different feeding techniques. In the spruce-fir forests of northern New England and Canada, five warbler species peacefully coexist in the same trees, each feeding in a slightly different manner: the Cape May Warbler looks for insects mainly in the tops of trees at the outer tips of the branches. The Yellow-rumped Warbler feeds nearer the trunk, on the lower branches of the tree, and on the ground. These two species also catch many insects in the air by flying from tree to tree. The Black-throated Green Warbler feeds primarily at middle elevations in the trees, on the tips of the branches and midway from the tips to the trunk. It hovers to reach insects more than any of the others. The Blackburnian Warbler searches the outer tips of the tree from mid-height to the top. The Bay-breasted Warbler feeds in the lower half of the same trees, mainly away from the outer tips of the branches. By dividing the tree into different sections (which may have different kinds of insects), five species are able to exploit what would seem like only a single food source.

Habitats may also be partitioned by the sexes of a single species: male Brown-headed Nuthatches usually forage lower on trees than do females, while in breeding Indigo Buntings the male forages high in the trees and the female rarely above 10 feet. The Huia, an extinct New Zealand passerine illustrated overleaf, had evolved differently

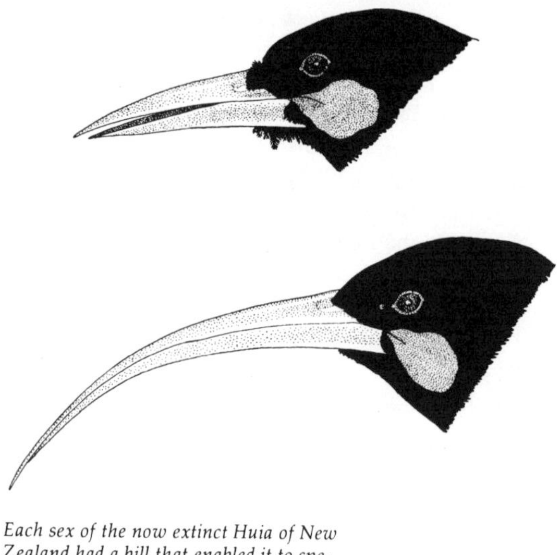

Each sex of the now extinct Huia of New Zealand had a bill that enabled it to specialize in different types of insect prey: the male, above, used its stout bill to chop rotten bark and wood, and the female, below, probed in narrow crevices the male could not reach.

shaped bills in each sex; the male's bill was short, thick, and relatively straight, used to dig into rotted wood for boring grubs, while the female's was long, slender, and strongly downcurved for probing into crevices.

Plant Eaters

Birds use many parts of plants for food. The most frequent are the buds, leaves, fruits, and seeds.

Swans, geese, and many ducks graze on the stems, tubers, and seeds of grassy plants, both above and under water. A Mallard or Black Duck tipping up is reaching plants growing below the surface. The longer necks of geese and swans enable them to reach deeper plants.

The buds and leaves of trees, evergreen and broadleafed, are staple foods of grouse and ptarmigan, which also take some bark and twigs. In spring you can often see grosbeaks and Purple Finches picking off buds high in the trees.

Many birds include seeds in their diet, but only a few specialize in them; some of the pigeons and doves and certain finches eat little or nothing but seeds. One theory explaining why most seed eaters are in the families considered to be most recently evolved (the finches and weaver finches) is that seeds — small, dry fruits scattered by the wind — are a recent evolution in plants. The only birds that could take advantage of the many new niches created by this newly available food source were the recently evolved, still very adaptable groups like the finches and weaver finches. The large number of species in these families is due to the absence of competition from other families for the new niches. Of course, it should be added that many of the seed specialists, like crossbills, redpolls, goldfinches, and siskins, sometimes placed in a family separate from the sparrows, feed on tree seeds — in cones and the catkins of birch and alder — which are not a relatively recent evolution of plants.

A few birds with strong bills, like jays, nutcrackers, and woodpeckers, hammer nuts open. The now extinct Passenger Pigeon, whose vast abundance was in part due to the plentiful nut crops in the eastern hardwood forests, had too weak a bill to break nuts open and instead swallowed them whole, grinding them up in the gizzard.

Fruit is not regularly available enough of the year in temperate regions like North America for any birds to make it the main part of their diet. Of North American birds, Cedar Waxwings eat the most, but many thrushes, orioles, and other insect eaters take fruit when it is abundant in late summer and fall. Robins are sometimes pests around cherry orchards and berry farms.

In the tropics, warmth and a greater variety of trees make fruit available all year. Many birds, including pigeons, parrots, toucans, and tanagers live entirely or almost so on fruit, although, like plant-eating birds of temperate regions, many feed insects to their young for protein. Many tropical fruit trees depend for dissemination on birds that swallow their seeds while eating the fruit and excrete them later at new localities. Other seeds are spread by birds that simply drop them away from the tree.

Fish Eaters

Fish are the regular diet of various birds living near water. Many oceanic birds, however, especially petrels, shearwaters, and albatrosses, feed primarily on planktonic invertebrates, shrimp, and squid, which are often much more abundant and easier to catch than fish.

Several of the techniques used for catching fish are illustrated on page 69. Loons, grebes, cormorants, mergansers, and alcids dive underwater from the surface, locating a particular fish only after they have submerged. Other birds spot a fish while in the air and make a plunge dive to get it — terns, pelicans, gannets, and the Osprey catch their fish this way. Kingfishers usually spot the fish they will plunge for while watching from a perch. Smaller birds naturally take smaller fish, and of the plunge divers only the gannet regularly goes far below the surface in pursuit of fish — it dives from as much as 100 feet in the air and swims at 50 feet beneath the surface.

Herons and egrets stand quietly at the edge of the water, waiting for a fish to swim by. With a sudden jab of their long bill they grab or occasionally spear the fish. Other fish eaters manage to avoid contact with the water entirely; frigatebirds and some tropical terns that lack waterproof feathers swoop to the surface of the water and pick up fish without ever landing, or catch fish and squid jumping out of the water to avoid larger fish below.

Piracy is a common practice among certain fish eaters. Jaegers and skuas catch their own food — small mammals, birds, and insects — while nesting, but spend the winter at sea chasing smaller birds, mainly terns, to make them drop the fish they have caught. Frigatebirds often use the same scheme, especially in windless weather when it is more difficult for them to swoop to the surface of the water to catch their own fish; they are such agile fliers that they usually catch the dropped fish before it reaches the water. Bald Eagles, primarily fish eaters, steal many from Ospreys; the eagle catches some for itself and also eats many it finds dead.

All birds that eat fish in one gulp swallow it head first; as the stiff fins and spines point back, it slides down more easily that way.

Predators

Anything that eats another animal is a predator, but the term is usually saved for those that take vertebrate prey. Some herons, ibises, and other larger shore feeders take frogs and snakes as well as fish, but nearly all eaters of birds and mammals are in the hawk or owl families. The term "raptor" is used to mean any hawk or owl.

Hawks and owls have long had a bad reputation as killers of chickens and game birds, but every study has shown that the bird in question feeds almost exclusively on mice and rats, which can cause extensive damage to crops, and rarely if ever takes what man wants. All hawks and owls should be fully protected, since they prey on animals whose populations would otherwise go unchecked, eliminating all the vegetation they could consume. Raptors, like other predators, usually take the weakest, sickly, least alert members of their prey species, thereby improving its genetic stock. All predators play an essential role in the natural scheme of things; where predators have been eliminated there is always a rise in the number of diseased or starving members of their prey species.

Hawks have evolved in different directions to exploit all food sources available. Buteos, including the Red-shouldered, Red-tailed, Broad-winged, and Swainson's Hawks, usually locate their prey while soaring or watching from a perch. They all take rodents, but the Red-shouldered, which usually lives in wet woodlands, also eats many frogs, snakes, and crayfish, and the Broad-winged catches many insects.

Different fishing techniques are illustrated by this imaginary gathering of birds from different oceans: a tern, Brown Pelican, Black Skimmer, Gannet, gull, Black-footed Penguin, and cormorant.

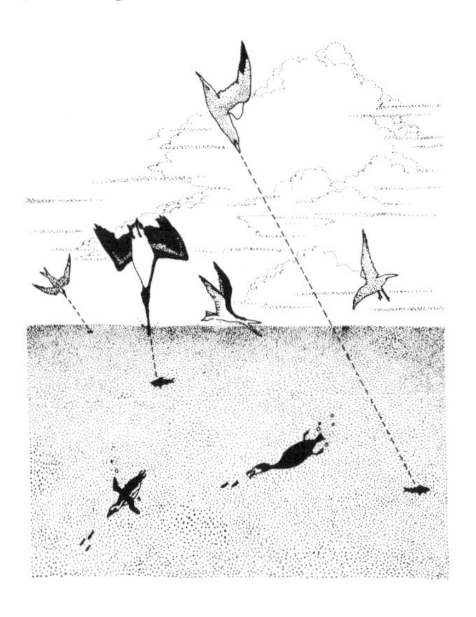

The Accipiters, or true hawks, feed primarily on birds, which they pursue in rapid chases through the woods. Each of the three North American species takes birds of different sizes, but the overlap in prey between the small Sharp-shinned Hawk and medium-sized Cooper's Hawk is great enough so that the two will not nest in the same woods. The Goshawk also takes many chipmunks, squirrels, and other woodland mammals.

The Golden Eagle has a varied diet. Large rodents like squirrels and woodchucks, rabbits, and birds the size of ducks and grouse are the main elements of its diet, but it is strong enough to kill raccoons, foxes, and even deer weakened by starvation in winter.

The Marsh Hawk is also an adaptable feeder and takes all the small mammals, birds, frogs, snakes, fish, and large insects that occur in its habitat of marshes and wet meadows. Its regular feeding method is to fly a few feet over the ground, alternately flapping and gliding, and pounce on any animal it surprises.

Falcons of various sizes inhabit all of North America. The large Gyrfalcon lives in the tundra of the far north and takes whatever is abundant, from the small mouselike lemmings to large hares, ptarmigans, and nesting seabirds. Medium-sized falcons like the Peregrine and Prairie Falcon feed on birds up to duck size and small mammals. When chasing a bird, the Peregrine will sometimes dive on it at tremendous speed, stun it with a blow from its clenched talons, and then pick it out of the air before it hits the ground. The smaller falcons, the Merlin and American Kestrel, feed on small birds, small rodents, and large insects.

An unusual hunting technique is reported for some forest-falcons of Central and South America: from a concealed perch, a forest-falcon will make a series of high call notes that attract small birds to investigate. These birds find the caller's exact location hard to place, and come closer and closer until the forest-falcon suddenly flies from its perch to catch one of them.

Owls take advantage of the many small animals active at night. Larger species like the Great Horned and Barred Owls often catch smaller owls and birds they find sleeping. Most owls live mainly on rodents, although smaller ones take many insects as well. Some are active during daylight (owls of the far north have to be, since during the summer there is little or no darkness); those that hunt at night often locate their prey by sound, and can hunt successfully in total darkness, where their vision is no better than that of humans. After eating, owls cough up pellets of undigestible fur and bones. The pellets are gray and most easily found in the woods in winter on top of snow; with practice, you can tell which species of owl produced a particular pellet on the basis of size, contents, and location.

In nearly every raptor species the female is measurably larger than the male, although plumages are usually the same. The size difference may reduce competition for food, or enable the pair to bring a wider range of food items to the young, but there are no satisfactory explanations as to why the female should be the larger. This phenomenon evolved at least twice, in hawks and in owls, suggesting it has some importance.

Scavengers

Nearly every fish- and meat-eating bird will feed on a dead animal. One reason for the bad reputation, noted above, of certain hawks is the assumption that a hawk found eating a dead lamb, chicken, or game bird was the cause of its death.

True scavengers, like the separate vulture groups of the Old and New World, feed almost exclusively on dead and decaying animal bodies. Their bills and feet are generally too weak to capture and kill live prey, or to tear apart bodies that have not already begun to decay. Vultures locate carrion while soaring high in the air. They space themselves so that each soaring vulture can see several others in various directions. Each watches to see if one of the others descends to food it has found, then flies over to join it; vultures still farther away will fly toward the bird that has noticed another descending. In this way a group of vultures can locate food far beyond the search area of any individual.

Scavengers are an important part of any wildlife community. Gulls and shorebirds scavenging on the beach, vultures, crows, and raptors inland, all help to eliminate health hazards and dispose of matter which would otherwise not be recycled into the plant and animal community.

Specialized Diets

Some birds have very specialized diets which, to a certain extent, free them from competition with other birds and animals. Hummingbirds live primarily on nectar from flowers. Many favor orange and red tubular flowers whose color and shape do not attract insects. In the tropics, some flowers depend exclusively on hummingbirds for pollination. Their diet of nectar restricts the size of hummingbirds, for no bird larger than a few inches could live on a diet of nearly all sugar. It has been calculated that a hummingbird much larger than the largest existing species, the 20-gram Giant Hummingbird of the Andes and Argentina, would not have enough foraging time to feed itself each day, and one smaller than the smallest existing species, the 2-gram Cuban Bee Hummingbird, could not store enough food to survive the night. Some larger hummingbirds, in fact, feed extensively on insects; even the smaller ones take insects from flowers and feed them to their young.

In the American tropics honeycreepers also feed from flowers. In Hawaii a separate family called the Hawaiian honeycreepers feeds on nectar, and several Old World families, especially the sunbirds and honeyeaters, fill the niche in the tropics of Africa, Asia, and the Australian region. All these birds are slightly larger than most hummingbirds and feed from a perch rather than hovering in front of the flowers.

Tree sap is the specialty of the sapsuckers and a few other woodpeckers. The Yellow-bellied Sapsucker makes several different types of holes in trees, depending on the season, type of tree, and weather; each type of hole produces sap most efficiently under certain conditions. Sapsuckers also eat some of the nutritious inner bark, fruit, and insects. A few other birds, including hummingbirds, will come to the holes made by sapsuckers to take a little sap or pick out the insects it attracts.

Wax is a major part of the diet of honeyguides, a family of 11 species found in Africa and Asia. No other birds or any other vertebrates eat wax, but the honeyguides have special bacteria in their digestive system to break it down. One species, the Greater Honeyguide of Africa, has developed the habit of leading men and honey badgers to beehives. The bird flutters and calls in front of the man or honey badger and will lead him or it to a hive where it waits for its helper to break the hive open for the honey. The bird then comes down to eat the exposed wax, showing very little interest in the honey or bee larvae. How this relationship between bird and hive robber developed is unknown, but today in Africa where people now get their sweets from a store and no longer follow the honeyguide, the bird has lost the habit of leading people. It is no less common than previously, and presumably still leads the honey badger.

Blood is a regular item in the diet of the Sharp-billed Ground-finch on Wenman Island in the Galapagos. The finch hops on the backs of Masked and Red-footed Boobies, common nesters there, and bites the booby on the rear part of the wing, beyond the reach of the booby's bill. The finch then sucks up blood from the bite. The Sharp-billed Ground-finch has not developed this habit on any other Galapagos island where it occurs, nor does any other bird so regularly feed on blood. This unusual feeding habit may be a learned, rather than inherited, ability, just as Great Tits in Britain have learned to pry off the tops on milk bottles, a habit that has now spread to the Continent.

Water

In addition to food, all birds require water. They can get it in three ways: by drinking, in their food, and by an internal chemical process combining oxygen with organic compounds containing hydrogen. No bird can survive very long on water produced by this last method, although, predictably, desert birds can survive longest. Many birds get all the water they need from their food, but others, including the desert birds whose diet is primarily dry seeds, must drink regularly and will fly several miles daily to a water source.

The ability to drink salt water is an essential adaptation for seabirds; failure to develop this ability may have restricted certain salt water birds to a coastal existence. Except for the kittiwakes, the only truly seagoing gulls, all gulls require fresh water, and can be seen drinking and bathing at reservoirs and lakes near the coast. Alcids and some penguins can drink both salt and fresh water, but petrels and shearwaters require salt water and will die of thirst when only fresh water is available.

Most young birds get all the water they need from their food until they are able to reach a water source themselves. Nestling sandgrouse in African desert regions, however, are brought water by their parents — the adult males soak up water in specially constructed abdominal feathers and fly back to the nest, often many miles from the water hole, for the young to drink it.

Most birds drink by sipping a drop of water, tilting the head back, and letting it slide down the throat. Only sandgrouse and pigeons are able to suck up water without tilting their head back.

Structural Adaptations for Feeding

To exploit such a variety of food items, birds have evolved many structural adaptations. Naturally, birds that must fly a lot to catch or locate their food have well developed wings and flight muscles; those that get their food on the ground have strong legs. By looking carefully at a bird you can often predict what kind of food it eats.

Bills

Animals without paws or hands have two ways of catching and holding their food — with mouth or feet. The majority of birds use their bill. Just as the finches on the Galapagos evolved differently shaped bills to handle different kinds of food, so have birds as a whole evolved a tremendous range of bill types, some perfectly adapted for a very specialized kind of feeding, others well suited to birds that take several kinds of food.

The bill is a horny sheath covering the bones of the upper and lower jaws, the mandibles. Nostrils are two small holes in the upper mandible. In nearly all birds, unlike mammals, both upper and lower jaws can move.

Among the passerines, there are four widespread bill shapes, shown opposite. Seed eaters, like grosbeaks and finches, have short, thick bills that can crush a hard seed. Foliage gleaners — warblers, vireos, orioles — have longer, thin bills that can reach farther to pick an insect off a leaf; as their food is soft, the bills don't have to be very strong. Ground probers like the Starling have a pointed, thin bill that goes into the ground easily. Birds that take flying insects — swallows, flycatchers, the American Redstart, and nonpasserines like swifts and nighthawks — all have flat bills with a broad base, the surest way to catch a moving insect.

A few passerine species have very specialized bills: the crossbills, for example, feed on the seeds in evergreen cones, and the crossed tips of their bills are used like wedges to spread the overlapping scales of the cones and get the seeds.

Within one family, the sandpipers probably show the greatest variety of bill shapes, ranging from relatively short, straight ones for

Different passerine bill shapes: from left to right, the Magnolia Warbler's unspecialized bill picks soft insects off foliage, the Evening Grosbeak's thick bill crushes seeds, the Starling's long, thin bill probes the ground, and the Barn Swallow's wide gape makes scooping insects out of the air easier.

picking food off the surface of the ground, to long curved ones for probing far underneath. Many shorebird species can feed together at a mud flat because each takes a different kind of food. Even among the most closely related species, like the Semipalmated and Western Sandpipers, bill shape is slightly different (although there is some overlap) and diet may be also. Curlews use their long, downcurved bills to reach deep into water for mollusks and small crabs and to probe in the mud for worms and insects. Godwits, with long, upturned bills, are more exclusively probers, as are the straight-billed dowitchers, whose rapid head jabs in the mud give rise to comparison with sewing machines. The American Woodcock uses its 2½- to 3-inch-long bill to get worms out of the ground. The tip of its upper mandible is flexible, and when underground can move away from the lower mandible and then close on a worm like a forceps.

Raptors have sharply hooked bills for tearing apart animal flesh. Herons and egrets have long bills which, with their long necks, enable them to make sudden, long jabs into the water for fish, frogs, crayfish, and snakes. Fish are sometimes speared; with a toss of its head the heron then shakes the fish loose and catches it between the two mandibles. Ibises use their long bills to search actively underwater for prey, rather than waiting like the herons.

The Brown Pelican does not use the pouch on its lower mandible to hold fish after they are caught, as is commonly assumed, but widens it underwater to enclose a fish, which is trapped when the upper mandible shuts. The operation is shown on pages 78 and 79.

The Black Skimmer feeds by flying low over the water, cutting the surface with its longer lower mandible. When it touches a fish or other food item, the bill snaps shut on it. Unlike other birds, the skimmer has a fixed lower jaw, attached to the skull. If the lower jaw were moveable, it wouldn't stay in place as the bird skimmed the water, and might actually be wrenched off.

The bright colors and patterns on the bills of some birds play a role in courtship and feeding young. Often these special marks disappear during the season they have no function, as the colorful stripes on the puffin fade in fall. The long bill of the toucans is spongy and lightweight; the bright colors may help in species recognition or courtship, but no satisfactory theory for its evolution to such a large size has yet been put forth.

Feet

Feet serve feeding birds in two ways: in getting to food, or in catching and holding it. The legs and feet are covered with scales, and each toe has a claw at the tip. Because the arrangement of the scales (scutellation) is always the same on all members of any species and is often similar in closely related species, genera, and families, it is sometimes used as a taxonomic characteristic. Other characteristics like the length of the toe or claw, presence of webbing, etc., that vary

Different bill shapes enable each of these shorebirds to take different food items from the same mud flat. From left to right, the Whimbrel, Hudsonian Godwit, Short-billed Dowitcher, Black-bellied Plover, Least Sandpiper, Semipalmated Plover, and Ruddy Turnstone.

within a family, are often adaptations related to feeding or locomotion.

Many marsh birds have long toes to support them on mud. As shown on page 90, the jacanas have toes and claws long enough to support their weight on floating lily pads, where they pick off insects. The long hind claws on larks and finches of sandy or snowy habitats similarly help support them.

Nearly all swimming birds use their feet, although some ducks and the penguins "fly" underwater with their wings. Different types of feet adapted for swimming are illustrated on page 80. Loons and grebes paddle with both feet at once; ducks swim with alternate strokes of the left and right foot. Most swimmers have webs joining the toes to create one flat paddle, but the toes of grebes and coots are broadly lobed.

Most raptors seize prey in their feet, which have sharp talons to help them kill and carry it. Ospreys have rough, scaly soles which, with their sharp claws, help grasp a slippery fish.

Parrots, which often eat fruit too large to swallow at once, pick it with their bill and then hold it in one foot while eating small pieces. Other birds, like the towhees, scratch away leaves and earth with their feet to get at food underneath.

Tool Use

A few birds have developed the ability to use tools to help them find or reach food. It is not known how these habits developed, nor

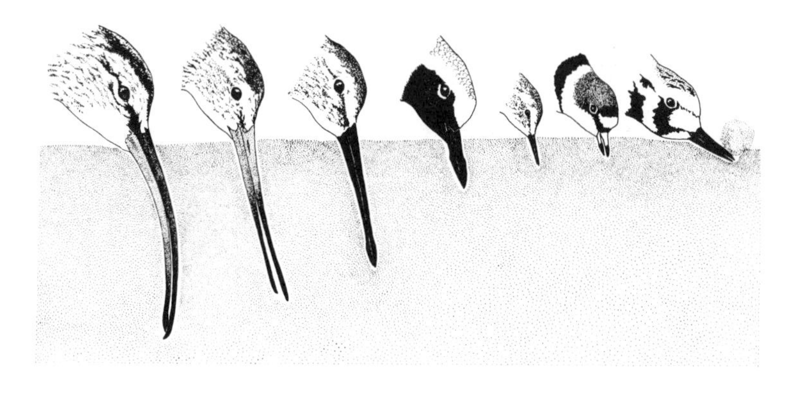

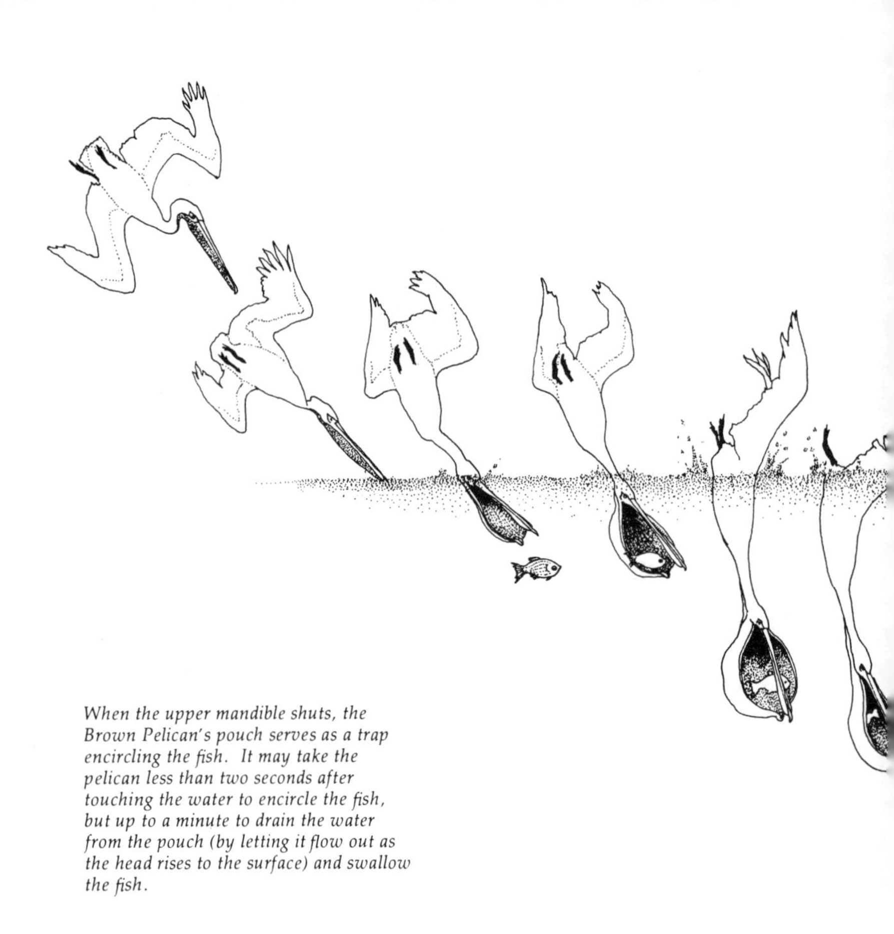

When the upper mandible shuts, the Brown Pelican's pouch serves as a trap encircling the fish. It may take the pelican less than two seconds after touching the water to encircle the fish, but up to a minute to drain the water from the pouch (by letting it flow out as the head rises to the surface) and swallow the fish.

whether they are now entirely inherited behavior or must be learned by the young.

One of the finches on the Galapagos Islands, the Woodpecker Finch, uses a twig or thorn to flush insects out of cracks in bark or holes it has made. When an insect comes out, the bird drops its tool to catch the insect. The technique is shown on page 81.

In East Africa, the Egyptian Vulture breaks open Ostrich eggs by throwing stones at them with its bill.

Almost in the category of tool using is the habit of gulls, crows, and Ravens to carry hard shellfish high in the air to drop on the ground. Gulls, however, often do not realize that the shell must be

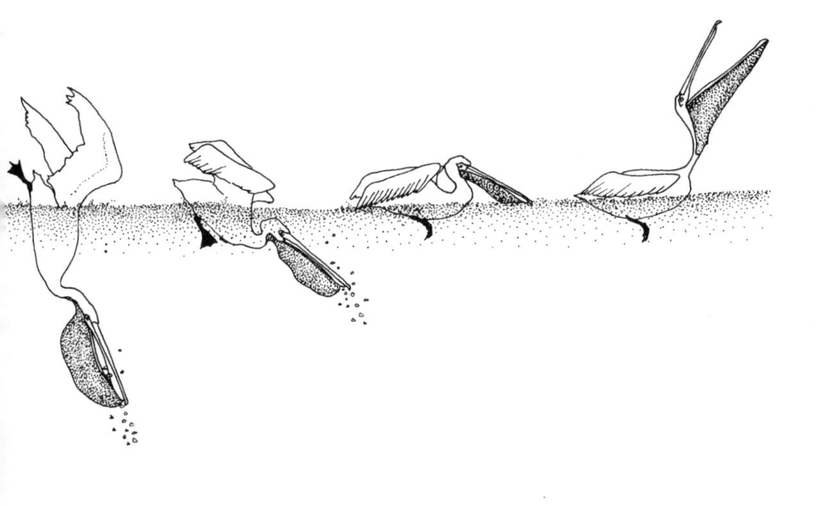

dropped on something hard to break, and some will repeatedly drop a shell on a sandy beach, never getting the food inside.

Group and Cooperative Feeding

Some birds feed in groups, which may be composed of one species or several. Scavengers usually feed in groups; groups of mixed species picking off one carcass are regular sights in Africa, where there are many vultures, and in oceans with a variety of albatrosses, shearwaters, and petrels. However, a group of birds in which each individual is after a separate food item is actually a more widespread form; flocks of blackbirds, Brown-headed Cowbirds, and Starlings are a frequent sight on fields in the fall: they are searching for insects and seeds. Usually they move in one direction, with the rear birds continually flying over the others, landing in front to search a new area; the flock seems to "roll" across the field. As further described in Chapter 11, small groups of chickadees, nuthatches, creepers, kinglets, and perhaps a few others move together through the woods in winter. Each species feeds in its own regular fashion; traveling together the group may more easily spot possible enemies. In the tropics similar mixed groups are even larger and more widespread.

Some birds associate with others for the food they stir up or discard. Phalaropes have been observed following Northern Shovelers, which vigorously paddle the water while feeding; all the paddling brings insect larvae to the surface. Birds will also follow cattle, as the Cattle Egret and Brown-headed Cowbird do, or farm machinery, as gulls follow tractors, for the insects they flush.

A few species assemble in flocks to feed in a manner that is more successful than individual effort. When Double-crested Cormorants find a school of fish, many may gather in a curving line, swimming forward and diving underneath after fish — this may limit the ability of fish to dodge. White Pelicans use a similar formation shown on

Different types of feet adapted to swimming: the lobed toes of the Western Grebe are typical of the entire family; the Brown Pelican, like all pelicaniform birds — tropicbirds, pelicans, boobies, cormorants, anhingas, and frigatebirds — has all four toes joined by webbing; in all swans, geese, and ducks, such as this Mallard, only the three front toes are joined.

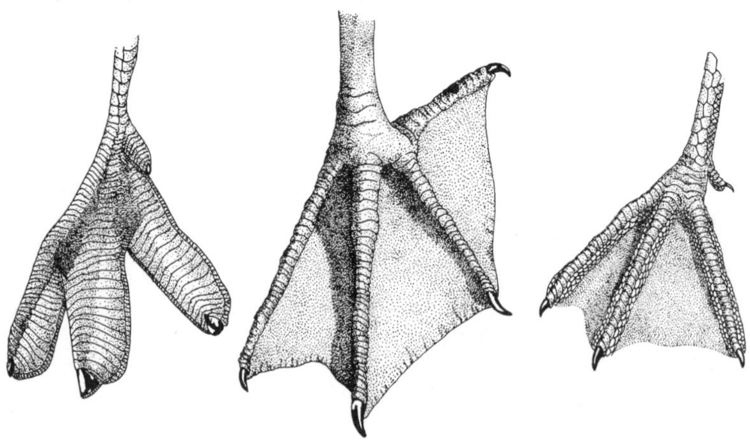

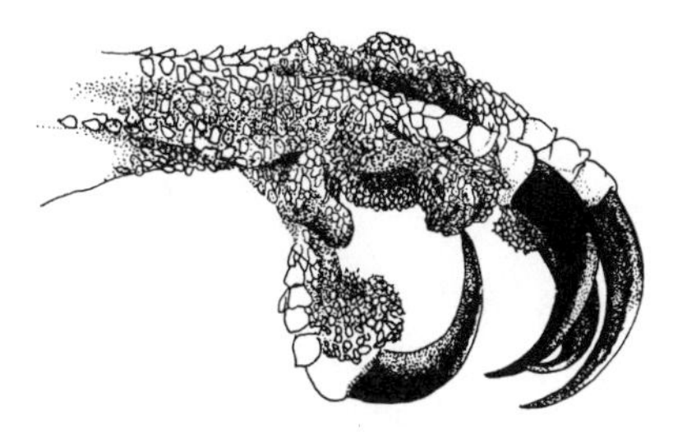

Rough skin and sharp talons help the Osprey hold onto the most slippery prey; in addition, the outer front toe is reversible, enabling the Osprey to grasp more evenly, with two toes on each side of the fish.

page 82; they gather on the water to drive a school of fish into the shallows where they are easily scooped up.

Food Storage

Food storage is practiced by several unrelated types of birds. A few predatory birds make temporary stockpiles for short-term storage. Barn Owls sometimes pile extra dead mice or rats near the nest when the young are not hungry enough to eat them. Both North American shrikes impale insects, reptiles, small birds, and small mammals on thorns, barbed wire, or in branch forks to eat later; the Loggerhead Shrike, in fact, rarely nests where some type of thorny or sharp-twigged tree does not occur.

In fall and winter, chickadees, nuthatches, woodpeckers, and jays hide seeds and nuts in tree crevices and in the ground. Some species and individuals remember the storage places better than others: jays and the related nutcrackers are responsible for many oak plantings,

Tool use by the Galapagos Islands Woodpecker Finch.

Groups of White Pelicans drive fish into the shallows by beating their wings as they swim in an encircling line toward reed beds.

which grow from buried acorns that were never recovered. Nutcrackers have the best memory of all the nut buriers; they can find items underneath the snow that were buried several months earlier.

The most elaborate type of food storage is practiced by the Acorn Woodpecker of the West Coast and Southwest, which bores long series of individual holes in certain trees, each to be filled with an acorn. Some insects are also stored in specially carved holes. The birds live in family groups of two to ten individuals, defending their storage and feeding trees against all other birds and mammals. Each group may have thousands of stored acorns.

Digestion

Because birds use up energy quickly, they have evolved a digestive system that rapidly converts food and gets it into the bloodstream. A quick digestive cycle also prevents birds, which need to be as light as possible when they fly, from carrying the extra weight of partially digested food. Fruit, easily digested, often passes through the system in half an hour, and ducks can digest and excrete shellfish and crayfish in a half hour to 45 minutes. Some seeds and nuts, which have to be crushed and ground up internally, may take a few hours.

A typical digestive system is shown below. Most birds have few taste buds and poorly developed salivary glands, if any. They swallow their food quickly. From the mouth, the food passes down the esophagus, or gullet. Some birds, especially eaters of seeds, buds, and leaves, like pheasants and grouse, eat food very rapidly, faster than it can be passed through the digestive system, and have a pouch-like crop, where food is kept to be digested later when the bird is in a safer place, resting or roosting. Other birds, especially those like hawks, vultures, and fish eaters that take large food items, have an esophagus that can stretch to hold food until it is brought back to the nest and disgorged to the young, or digested by the adult.

The stomach has two parts, a proventriculus with gastric juices that begin breaking down the food, and a ventriculus, or gizzard, which crushes hard food items. Birds that eat soft insects or fruit have a small gizzard because their food can be broken down easily

The digestive system of a chicken.

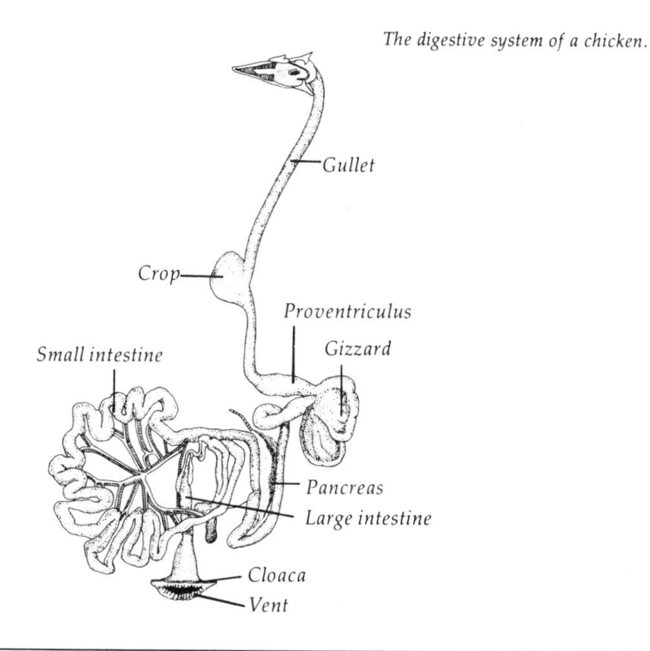

by gastric juices alone. Eaters of seeds, nuts, shellfish, or other hard items often have a small proventriculus because there is little that gastric juices can do to a shell or the hard coating of a seed or nut. Their gizzards are muscular, and break food items by contracting and crushing them. The evolution of internal means of breaking down food has enabled birds to do away with the heavy teeth and jaws found in fossil birds like *Archaeopteryx*. The location of the digestive processes closer to a bird's center of gravity gives it greater stability in flight.

Many birds swallow pieces of sand or gravel, which pass into the gizzard and help grind up hard food items — Mourning Doves are among the easiest birds to observe deliberately picking up sand or fine gravel at the beach or from fields and road edges. Shorebirds feeding on the beach or robins pulling worms out of loose soil pick up sand accidentally with their food. Other birds may eat sand as part of their diet for the minerals like calcium and phosphorus it contains.

After some breakdown in the proventriculus and/or gizzard, food items pass to the small intestine where they mix with bile from the liver and pancreatic enzymes from the pancreas. These chemicals continue the breakdown of food into carbohydrates, fats, and proteins. The length of the small intestine depends not only on the overall size of the bird, but also on its diet. Plant eaters, especially eaters of buds and leaves, must consume a much greater bulk of food to derive the same amount of nutrition. Their intestinal tract must be longer, to process the larger volume of food material. Intestinal lengths vary from the two inches of the Ruby-throated Hummingbird to the Ostrich's 46 feet.

The "large intestine" is actually very short in birds, since digestion is relatively complete by the time food passes through the small intestine. Between the two intestines are paired caeca, which in some birds contain bacteria that further break down food particles.

Excretion

The undigestible parts of food items that pass through the system are excreted through a vent at the cloaca, the end of the large intestine. Liquid wastes from digestion are carried by the blood stream to the kidneys. After process in the kidneys, the wastes are carried to the cloaca in the ureters. Useable water in these wastes is reabsorbed in the kidneys and ureters; what remains is excreted from the vent. The excretory system is shown with the reproductive organs in a drawing on page 107.

The kidneys alone cannot filter out all the excess salt in the diet of many oceanic and brackish-water birds. These species have special salt glands in the front of the skull to remove much of it; the resulting salty liquid oozes out of the nostrils. The tubes for getting rid of this salt solution are most visible in the albatrosses, shearwaters, and petrels, (the source of their group name, "tubenoses") but salt glands are found in all seabirds, ducks of marine and brackish environments, coastal feeders, and even sparrows like the Song, Savannah, and Sharp-tailed Sprrows that live in brackish marshes. Races of these sparrows that do not live in environments where their diet contains a lot of salt lack the salt glands.

We have seen how diet affects a bird's anatomy, internal and external. The bird's body is first a food-getting mechanism, and any modification that hampers its ability will not be tolerated by natural selection. The profound effects of diet on breeding behavior, social systems, migration, and distribution will be shown in later chapters.

New studies measuring the actual energy values of various bird foods are giving us a much more precise idea of what each bird extracts from the environment and how much it requires for each activity.

This Greater Shearwater shows the prominent tubular nostril, connected to the salt glands, typical of all albatrosses, shearwaters, and petrels.

Chapter Six

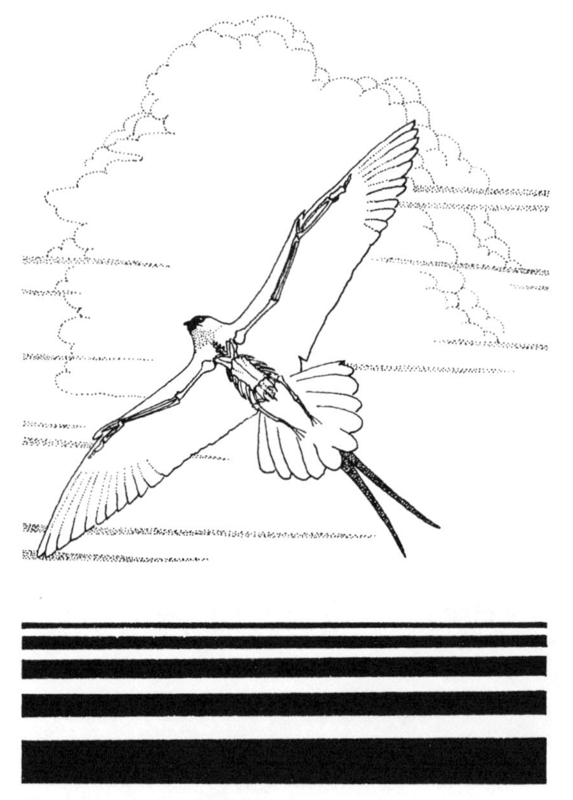

The skeleton of a Long-tailed Jaeger, seen against the outline of its wings extended in flight.

Anatomy

BIRDS OWE their success to the adaptations each species has evolved to survive the challenges of its environment. Like the external features of wings, tails, bills, and feet that we have already examined, the internal systems of a bird are shaped to serve specialized needs as efficiently as they can. The skeleton, muscles, heart, lungs, and other body parts evolve more slowly than external features; many can be used as taxonomic characteristics identifying all members of a family or order, but others may vary subtly from species to species.

The Skeleton

The skeleton holds the bird together, providing anchorage for the muscles and protection for the organs. The bones determine the overall size and shape of the body, and in this way respond to subtle environmental pressures. In species with a wide geographic range, individuals from the colder part of the range generally have slightly larger bodies than individuals from the warmer part, because heat is lost more slowly from a large body than from a small one of the same kind. For the same reason, individuals in colder parts of the range have shorter bills and other protruding parts, because the smaller the protruding parts, the less surface area there is to lose heat from.

A chicken skeleton that you may compare with a bird at home is shown on page 88. Flying birds have longer wing bones and most non-running birds have shorter legs, but basic features are the same for all birds. Compared with a mammal or reptile skeleton, there are many fewer bones. This is because flying animals need a light but rigid skeleton, which is most easily achieved by the fusion of some smaller bones and the elimination of others. Teeth, and the heavy

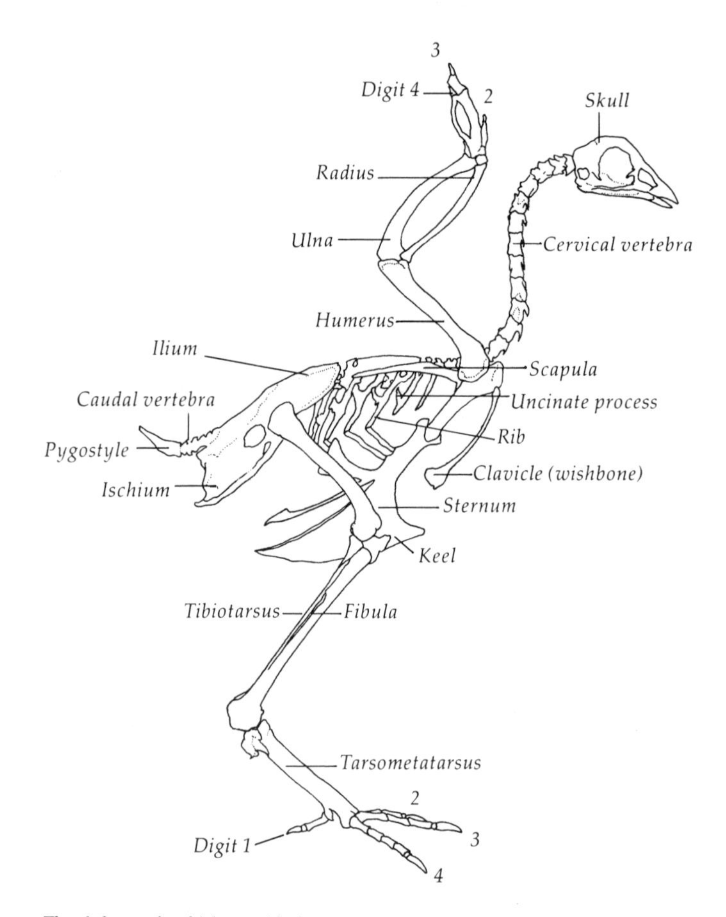

The skeleton of a chicken, with the principal bones labeled.

jaws to support them, are gone; so is the bony tail of reptile ancestors and *Archaeopteryx*. Many of the skull bones that are separate in reptiles are fused in birds.

Wings and legs have the same basic parts as our own limbs, but highly modified. While we have 29 bones in our forelimb, the pigeon has 11. Drawings opposite show the parallels between bird and human limbs. Among the differences, note the reduction of fingers and of bones in the wrist. In birds the equivalent of the first and fifth fingers are gone, and the second is fused to the third. The

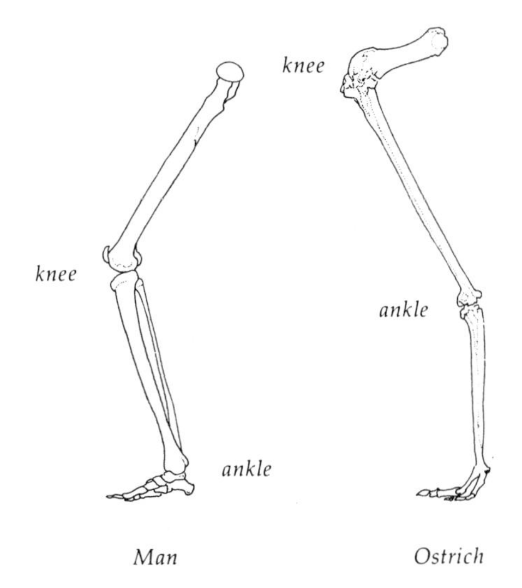

Human and Ostrich leg bones (not to scale): note how short the femur, the upper bone, has become in the bird, and how high the ankle is, so that the bird walks only on its toes.

Condor and human arm bones (not to scale): in the condor the first and fifth digits have been eliminated; primaries are attached to the third and fourth.

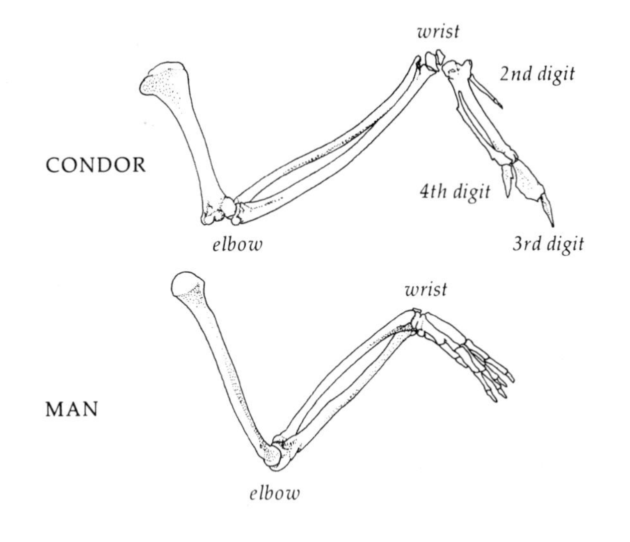

phalanges of the third and fourth digits hold the primaries. Many of the leg bones are fused. In contrast with the human foot, all of which is used for support, birds stand and walk only on the ball of their foot and with their toes. Most birds have four toes, arranged differently in various families for perching, swimming, climbing, or running. All passerines, and most other birds, have three toes pointing forward and one pointing back. Toucans, parrots, cuckoos, and most woodpeckers, all active climbers, have two pointing in each direction; some woodpeckers have lost one of the rear toes. Some running birds, like the Golden Plover and Sanderling, and alcids, which use their wings more than their feet to swim underwater, have only the three forward toes. The Ostrich has just two thick toes, which also point forward.

The sternum, or breastbone, is highly modified in birds. Three are shown on page 92. The Ostrich and other large flightless birds have a flat sternum, but all others have a keel-shaped sternum, which creates more surface area for the attachment of flight muscles. Fliers that beat their wings powerfully, like falcons, swallows, and hummingbirds, require more muscles than do soaring birds such as albatrosses and vultures or weak fliers like chickens, and consequently have a deeper keel. The sternum also protects the internal organs. At the front end, the sternum is connected to the shoulder girdle by two powerful coracoids, which attach with the clavicles (wishbone)

Toes: the Jacana's long toes and extremely long claws let it spread its weight on floating lily pads where it searches for food; the Hairy Woodpecker's arrangement of two toes pointing forward and two back gives added clinging support; the Solitary Vireo's foot shows the arrangement of all passerines — three toes forward, one back.

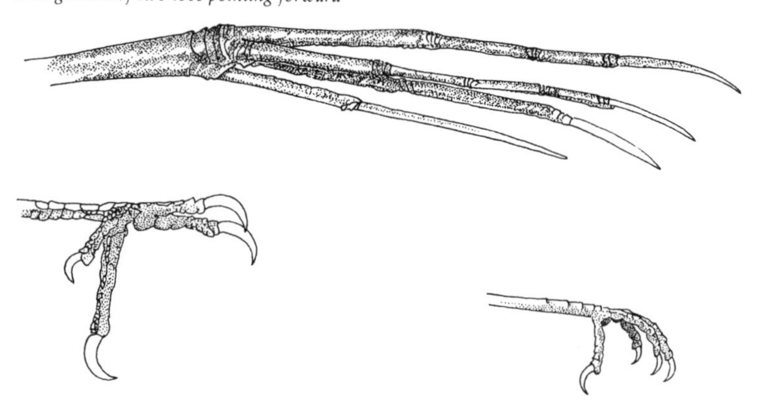

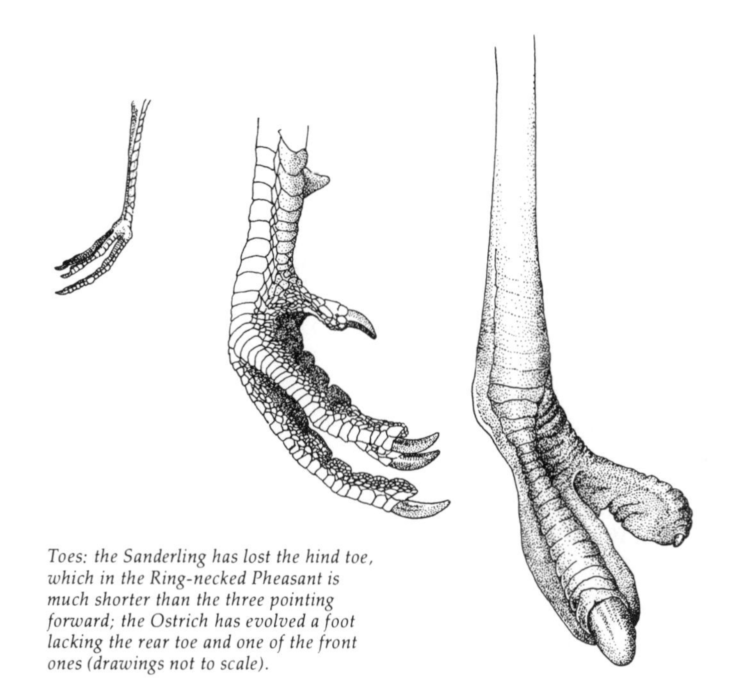

Toes: the Sanderling has lost the hind toe, which in the Ring-necked Pheasant is much shorter than the three pointing forward; the Ostrich has evolved a foot lacking the rear toe and one of the front ones (drawings not to scale).

and scapulae to create the strong support needed for powerful flight.

Bones of the rib cage connect the back with the sternum. Each rib is connected to the adjoining ribs by a small bone, all together called the uncinate process. Only birds have this system, which provides additional support.

An unusual feature of the bird skeleton is the cervical vertebrae, the neck bones. Unlike mammals, which all have seven cervicals no matter how long the neck, birds have cervicals that vary in number as well as size, from 13 in some passerines and cuckoos to 25 in swans. Most birds have 14 or 15. The flexibility of the neck enables a bird to stretch easily for feeding, preening, or looking around. This is an important adaptation, since fusion of many bones in the rest of the body has reduced its flexibility. Sometimes within one species individuals have one cervical more or less.

As a further adaptation for flight, bird bones are extremely lightweight. Many bones have air spaces, and the longer limb bones are sometimes completely hollow. The hollow bones have an internal support system of struts, as shown opposite. Because air is a good shock absorber and because bones are easier to bend when hollow than when solid, hollow bird bones are actually stronger than solid bones of the same weight. In some larger birds the humerus, the inner wing bone, contains an air sac extending from the lung. Birds for which lightness is no advantage, or an actual disadvantage, as in deep-diving loons, have solid bones.

The Muscular System

To perform their variety of movements, most birds have about 175 different muscles, usually in pairs. The muscular system of a bird is the part we eat. Looking at a cooking chicken, you will see that nearly every part of the body is covered with muscles, except the inside of the rib cage, where the heart, lungs, liver, kidneys, and other organs are found. Below the ankle joint are only the tendons, with no muscles and therefore little worth eating, so poultry is often sold without the feet. The "dark meat" is muscle made primarily of red fibers, which contain more capillaries supplying blood to the system and which store fat for sustained muscular action. These are the muscles used for activities requiring a lot of power — flying for most birds, which have dark breast muscles, and running for ground

Three types of sternum (not to scale): seen from the side, the Peregrine's has a large surface area for the attachment of muscles, while the weak-flying Bobwhite's has substantially less. The Ostrich's flat sternum is seen from the front.

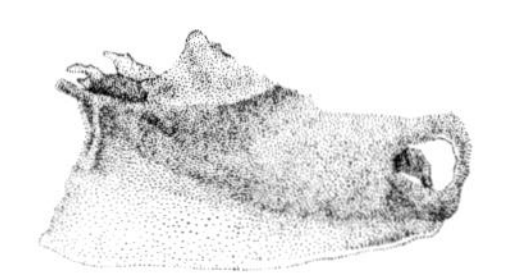

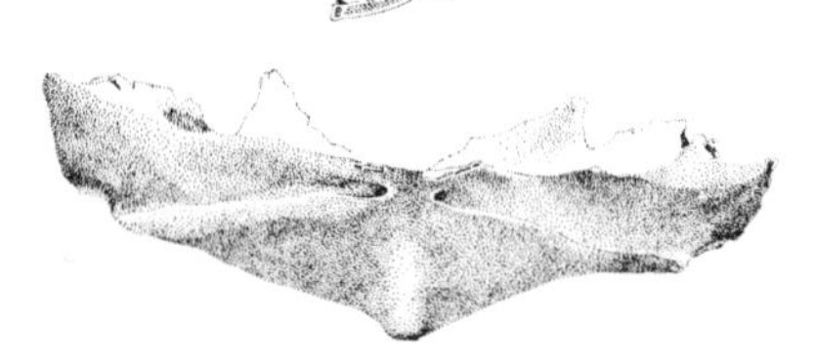

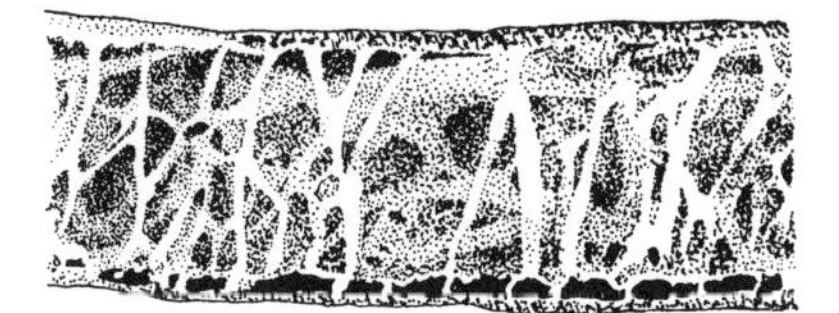

A cross section of a hollow bone showing the support system of struts.

birds like chickens and turkeys, which have dark leg muscles. The muscles made of white fibers have little fat and can only work for short periods of time; one of the reasons chickens cannot fly very far is because their breast muscles are of white fiber.

Many muscles are involved in the complicated motions of every up and down flap of the wing in flight. Different muscles are used to raise and lower the wing. The most important muscle for the powerful downward flap, which moves the bird forward through the air in addition to providing lift, is the *pectoralis major,* anchored at one end on the sternum and clavicle and at the other on the lower surface of the humerus. The main muscle raising the wing in flight is the *pectoralis minor,* found underneath the *pectoralis major.* It too begins on the sternum, but, as a tendon, runs through an opening in the shoulder bones to attach on the upper surface of the humerus. Like all muscles, it operates by contracting when stimulated by a nerve and then relaxing; when the *pectoralis minor* contracts, the wing is pulled up, working like a rope and pulley system.

Every feather is individually connected to a muscle, which enables the bird to adjust wing and tail feathers in flight, raise display or crest feathers, and erect or depress body feathers for bathing, cooling, or insulation.

Jaw muscles vary depending on feeding habits. Birds that crush seeds in the bill, like grosbeaks, have strong adductor muscles, which open and close the jaw. Shorebirds, Starlings, and other probers that open their bill underground to grasp something have large protractor, or gaping, muscles. Those birds that simply pick up and swallow their food, like many warblers, have no set of jaw muscles especially developed.

The size of leg muscles depends on how a bird uses its legs and feet. Active swimmers, runners, and climbers have stronger muscles than species that walk little or not at all. Many specially adapted muscles are involved in perching. Most important are two muscles of the lower leg (our calf), the *gastrocnemius* and *peroneus longus,* and the flexors, small muscles which become tendons and run the length of the foot down to the tip of each toe. When a passerine perches, the ankle joint is flexed, applying tension to these tendons and automatically locking the toes in a firm grasp of the branch.

The Circulatory System

As in our own bodies, the circulatory system transports nourishment to all parts of the body. The bird heart is divided into right and left halves, each half with two parts, an atrium and a ventricle. (Similarities to the human heart are a result of convergence, not common derivation.) As illustrated opposite, blood from the veins enters the right atrium, passes to the right ventricle and then to the lungs, where it discharges carbon dioxide and absorbs a fresh supply of oxygen. The blood then returns to the left atrium, passes through the left ventricle, and into the system of arteries directing blood throughout the body.

Birds that use energy very quickly, especially powerful fliers, have proportionately larger hearts than nonfliers, weak fliers, and soarers, all of which use less energy in flight. Birds of high altitudes and high latitudes also have proportionately larger hearts.

The heartbeat is extremely rapid in birds, especially in the smaller and more energetic species. The highest recorded rates are found in

The operation of PECTORALIS MINOR *to raise the wing: at left* PECTORALIS MAJOR *has just contracted, lowering the wing; as it relaxes and lengthens (center drawing),* P. MINOR *contracts, forcing the wing to move up, as shown at right.*

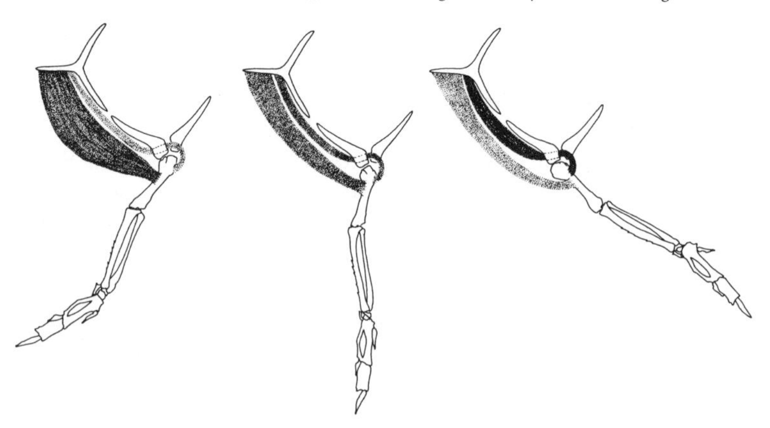

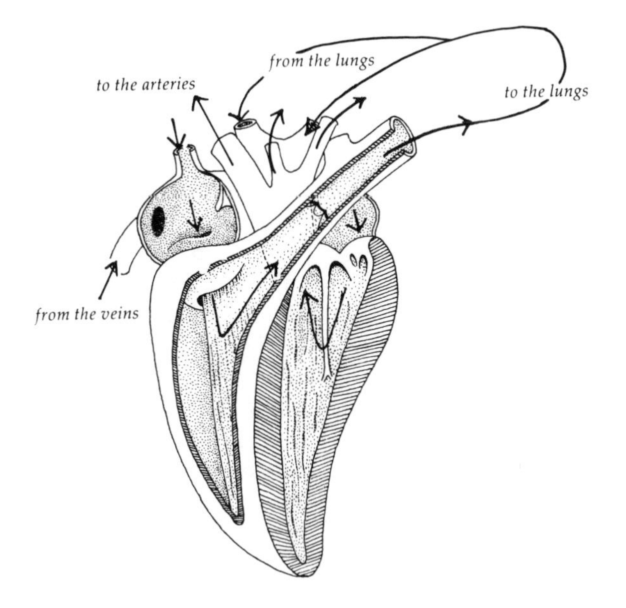

The route of blood through the heart (from D.A. Ede, BIRD STRUCTURE, *Hutchinson Publishing Group Ltd.).*

hummingbirds — when a hummingbird is extremely active the rate may exceed 1200 times per minute; under less stress, it may drop to 500. In large and less active birds the rate is considerably lower; for an Ostrich, it ranges between 38 and 176 beats per minute. (The human heart rate is 60 to 75.) In cold weather the heart rate is faster, to supply energy more rapidly to the body and maintain a high body temperature.

Although birds, like mammals, are called warm-blooded because their bodies maintain a steady temperature independent of the external environment, the temperature of a bird's body does vary within about 10° F under different conditions. The normal body temperature is between 100° and 112° F for different species, highest in passerines. Body temperature rises when a bird is active, feeding, flying, or digesting, or when the environmental temperature is high, and falls when birds are resting or asleep, or the surroundings are cold. Birds active in the daytime usually have the highest body temperatures around midday; for species active at night the rhythm is reversed.

Young birds, especially those born without a covering of down, have almost no ability to regulate their body temperature, and require brooding to protect them from extremes of cold and heat. With the development of feathers several days after hatching, young birds have enough insulation to maintain a steadier body temperature.

A few birds are able to lower their body temperature, heartbeat, and breathing rate during short cold periods, consuming practically no energy. Torpidity seems particularly an adaptation of birds whose food source may temporarily disappear during unseasonable cold spells. Some species of swifts, nighthawks, and swallows, all of which feed on flying insects that disappear during periods of cold weather, have been found dormant and mistaken for dead until they were warmed; they then flew away in perfect condition. Hummingbirds, which may not be able to maintain their high heart rate on cold nights, often become torpid until the air temperature rises the next morning. Because hummingbirds burn up energy so quickly, by lowering their heart rate (and therefore their body temperature — to about 70° F) at night they use less fuel from food consumed during the previous day, and avoid literally starving to death.

In extremely hot weather, birds have several means of maintaining a body temperature lower than that of the environment. They can become less active, rest in the shade if there is any, and depress their body feathers to reduce layers of insulation. Panting releases heat and moisture to the environment. Fluttering the gular membrane of the throat, which has many blood vessels, also helps the bird to lose heat and is especially used by birds with unfeathered gular membranes, such as pelicans, cormorants, and frigatebirds, which all nest in open sites where there is no shade. Other birds lose heat through the unfeathered legs and feet. Storks and New World Vultures defecate on their legs, which are cooled by the drying process; the sharing of this unusual habit is one of the reasons some ornithologists feel the two groups are related.

The Respiratory System

The bird's respiratory system shows several adaptations that permit rapid intake of large amounts of oxygen while fitting into a small streamlined body. Bird nostrils, or external nares, are small holes usually located near the base of the upper mandible. In most birds the two nostrils are separated by a thin wall of bone, but rails, cranes, and New World vultures lack the separation. In hawks and some parrots, including the common pet parakeet, the Budgerigar, the nares are surrounded by a soft membrane. As noted earlier, the "tubenoses" are so called because the nares are at the end of a tube on the upper mandible.

The respiratory system is illustrated on page 98. From the nares, air passes through a tube called the pharynx, past a slit-like opening, the glottis, into the trachea, or windpipe. The long and flexible trachea is made of a series of stiff rings that are bony on the lower surface, which is less protected by strong bones from possible bruises, and softer on the upper surface, where covered by the bones of the neck and sternum. At the end of the trachea, in the rib cage, are the two lungs, relatively small but divided into sets of increasingly fine membranes, about a thousand in each lung. By being divided into so many air passages, each exposed to blood vessels, the lungs are able to supply a large quantity of oxygen while occupying a small space.

To help supply heavy oxygen demands, the lungs are supplemented by a series of air sacs that let the bird inhale more air than the lungs can hold at one time. These air sacs fill all the available spaces within the body cavity and also contribute to the bird's lightness; the size and number of air sacs depend on how much buoyancy a bird requires for flight. Soaring species have air sacs extending to their hollow bones, others have a series of small air sacs paralleling the trachea. Some diving birds use the reserve air in the sacs while submerged, but deep divers have reduced air sacs to eliminate buoyancy.

External nares of various birds: the unseparated nares of a Sandhill Crane, the Peregrine's round nostrils surrounded by a soft membrane (cere), and the common pigeon's below the swollen sensitive skin called the operculum. See also the Greater Shearwater on page 85.

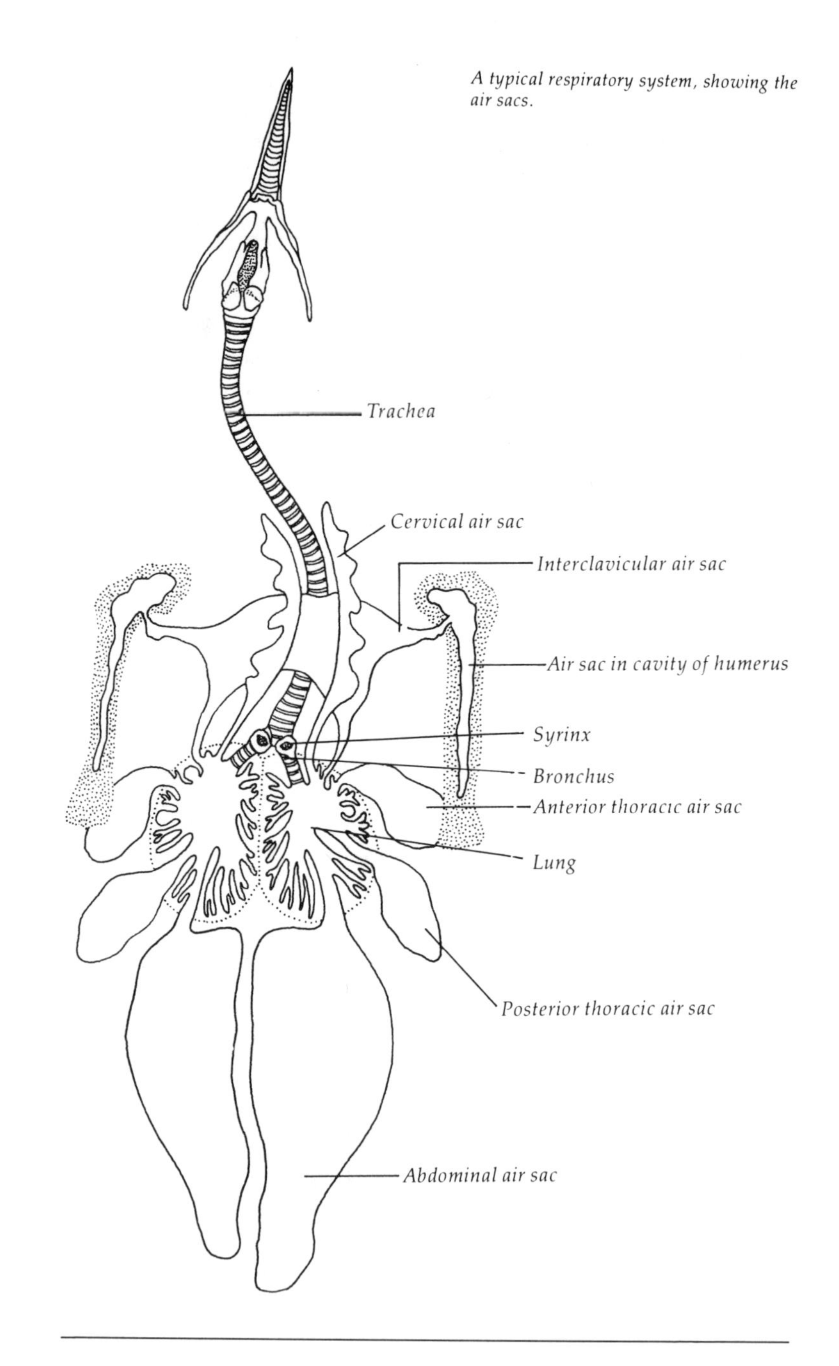

A typical respiratory system, showing the air sacs.

The route air takes through the lungs and air sacs is not well known. Some researchers believe air moves through the fine passages of the lungs and then into the air sacs, which act as bellows, forcing air back through the lungs and out to be exhaled. Others believe that most air goes directly to the air sacs through the larger lung passageways, then traveling through the finer passages of the lungs, with all gaseous exchange taking place as the air is on its way out of the lungs. In either system, the lungs are completely emptied with each breath. In flight, although not when a bird is at rest, the air sacs may also be emptied with each breath.

The breathing rate, like the heart rate, varies from species to species and with activities and conditions. Most small birds breathe between 100 and 200 times per minute, but faster when in flight or excited. Larger birds breathe more slowly — a resting Ostrich only 6 to 12 times per minute, but of course faster after exercise.

The Brain

The brain controls all voluntary movements of the body and receives and interprets experiences of the sensory organs. As in our own bodies, sights and sounds are perceived by the eye and ear, but only after the nervous system transmits the impressions to the brain are these sensations understood.

The size of various parts of a bird's brain reflects the relative importance of the areas they control. The optic lobes, related to vision, are large, while the olfactory lobes are small, since birds have a poorly developed sense of smell. The large cerebrum controls most of the instinctive behavior patterns. The cerebellum controls muscular coordination and, in part, balance. It is well developed, reflecting the many skills birds have and the precise movements they make.

Experiments with Canaries and Zebra Finches, two species that learn their song by hearing other individuals, have shown that areas of the brain controlling vocalizations are as much as five times larger in males than in females, which do not sing; this may also be the case in many other species where only one sex sings.

The brain is especially well developed in the parrots, owls, and woodpeckers, but is proportionately largest in the corvids — crows, ravens, jays, and magpies — a family always considered "intelligent" for the ability of various species to adapt quickly to changing circumstances. That the Common Crow population increased at the

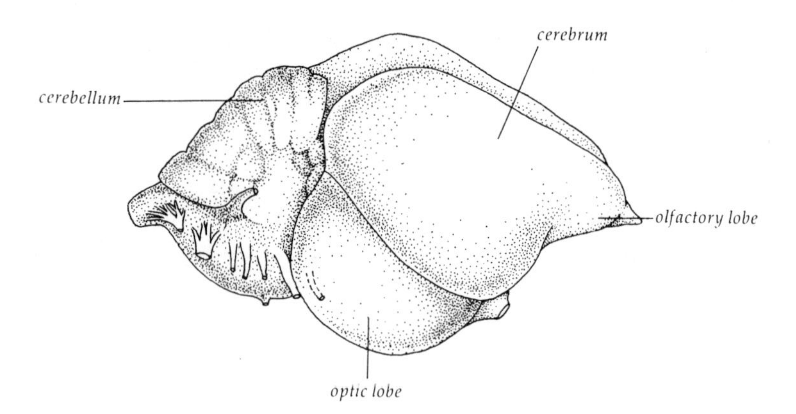

The brain of a COTURNIX *quail. (Reprinted by permission from* THE COTURNIX QUAIL *by Theodore C. Fitzgerald, © 1970 by Iowa State University Press, Ames, Iowa.)*

very time it was most heavily persecuted seems proof of the bird's resourcefulness. Like many corvids, it can distinguish a man holding a gun from one holding a stick, and after brief experience recognizes a former hunter even when he is unarmed.

Sense of Touch

Birds are so completely covered by feathers that there are few places where the skin is exposed to feel things. That birds often scratch themselves vigorously would suggest, however, that they feel minor irritations on the skin. The unfeathered portions of the legs and feet naturally feel the surfaces on which they walk, or the food and perches they grasp. To their great advantage, many birds do not seem to feel extremes of heat and cold on their feet — gulls and ducks may stand for hours on ice with no apparent discomfort.

The bill has sensory nerve endings especially developed in the birds that use it for probing or locating food they cannot see. The edges of duck bills have sensitive plates to sort food items out of the mass of vegetation and mud on the bottom of ponds. Bird tongues, instead of taste buds, have nerve endings that respond to texture. The bristles around the bill of certain birds have sensory cells at the base of the feather; these may be present to give clues about objects they grasp in the bill without seeing well, since many night-feeding birds have particularly large and profuse bristles.

Vision

Vision is the bird's most highly developed sense. Unlike most animals, they see in color. Nearly all birds find their food by sight, and many must be visually alert to avoid predators. Birds can distinguish objects much farther away than can humans, and their vision is in fact the most highly developed of any animal. Captive birds kept outside have often been observed looking nervously at the sky, watching a hawk that humans could detect only through binoculars. Similarly, the hawk in the sky can see and dive on small animals from a distance at which, to a human, the animal would be invisible.

The eyes of most birds are located on the side of the head, allowing them to see over a larger area than if both eyes faced forward as ours do. Most of the area a bird sees is perceived with only one eye; the fields of vision of the two eyes only overlap in a small area in front. (Penguins have no overlap at all; they always see two entirely separate images.) A disadvantage to perceiving an object through only one eye is the difficulty in judging distance, and for this reason you often see a bird cock its head to carefully focus on an object; likewise, the bobbing up and down of the head characteristic of many shorebirds may be an effort to gauge distances. Hawks have a greater overlapping range of forward vision, since they usually pursue prey in front of them, and owls, with their eyes in the front of the head, see only forward, the fields of both eyes overlapping almost entirely. The owl compensates for this limitation of field by being able, unlike other birds, to turn its head completely backward. At the other extreme is the American Woodcock, with eyes placed so that it can see in a complete circle all around its head. The woodcock is thus able to see predators approaching behind it, an important adaptation for a bird that often has its head bent over the ground while it probes for earthworms. The bittern's eyes are placed so that it can see forward when it "freezes" with bill pointed upward to blend in with the reeds of its marsh environment; when the bill is pointed forward, its normal position, a bittern can see food items directly below.

The eye of a bird is relatively large — the European Buzzard has eyes as large as man's. The two eyes of a bird may, in fact, weigh more than its brain. Beneath the surface, the area covered by the eye

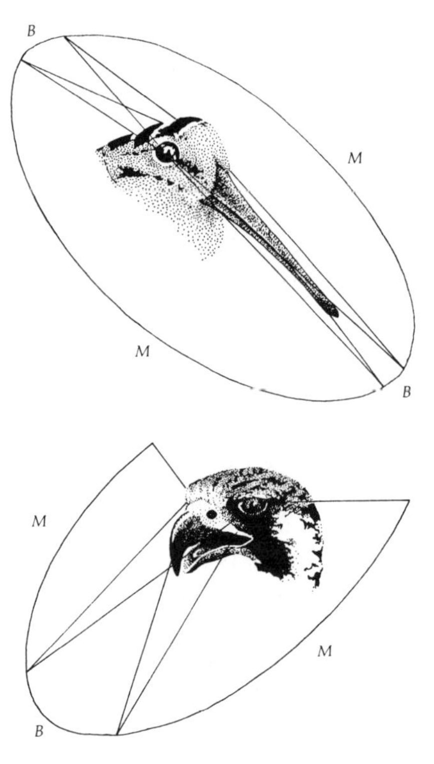

Ranges of monocular (M) and binocular (B) vision in the Great Horned Owl, American Woodcock, and Peregrine.

is much greater than what is visible externally. The eye's shape varies from group to group; in most passerines it is flattened, in raptors active by day the cornea, or outer surface of the eye, bulges, and in owls it is practically tubular. Bony plates of the sclerotic ring hold the eye in place and affect the amount of bulge in the cornea. Muscles control the curvature of the cornea and the shape of the soft lens beneath it, allowing the eye to quickly change its focus from far to near.

The rear wall of the eye is densely lined with the rods and cones that form the retina, the surface on which images are formed. The rods are sensitive to light, especially at low intensities; the more rods it has, the better a bird is able to see at dusk or at night. Species active at night predictably have more rods than do other birds. The cones function in bright light to form sharp images and distinguish shades of color. Among hawks and eagles, which are generally considered to have the keenest vision of all, the cones may be as dense as one million per square millimeter. (Man has only a fifth as many in the same area.) There is some evidence that hawks may have sacrificed some of their color vision to achieve sharper perception of form, distance, and motion.

The ratio of optic nerve fibers (going to the brain) to visual cells (rods and cones) is much higher in birds than in mammals, so that a bird's brain perceives the impressions of the retina with less blur or summation of fine detail than a mammal's does. In addition, for focusing sharply on an object, most birds have at least one sensitive spot, or fovea, on the retina. The fovea is a depression creating an increased surface area that permits an even higher concentration of cones. Birds like hawks and swallows, which pursue prey in front of them, have two foveae, one near the center of each eye, for spotting things, the other in the section of the retina that receives impressions from the area in front of the bird seen by both eyes at once.

The red, yellow, or orange oil droplets found on the cones in many birds' eyes probably serve as a filter, helping in the perception of various colors and cutting down the glare or brightness of others. Kingfishers, for example, have a very high proportion of red droplets, which may make visibility through the surface of water clearer.

The pecten is a small comb-like structure filled with tiny blood

The parts of a bird's eye.

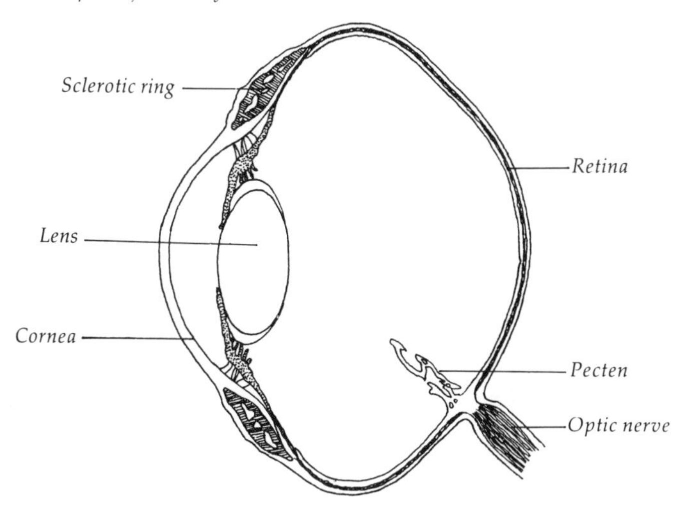

vessels on the surface of the retina. Its many blood vessels presumably help nourish the eye; several other functions of the pecten have been suggested, such as casting a shadow on the retina when a distant object is perceived, and thereby helping to detect movement. The most complex folds in pectens are found in hawks and insect eaters active by day; the simplest pectens occurs in birds active at night, indicating a function related to acuity in bright light. As with the oil droplets, our ideas about the pecten are still mainly theoretical.

A semitransparent third eyelid, the nictitating membrane, is an adaptation derived from reptiles, and found also in some mammals, including cats and camels. In birds the membrane cleanses the eyeball and may also protect the eye when a bird is flying into or facing the wind, or, for aquatic birds, when under water. Some birds blink the nictitating membrane frequently; you can see this from two or three feet away. In covering the eye, the membrane moves from the lower part of the corner near the bill upward to the opposite corner.

The eye color of most birds is dark, although there are many exceptions. Within the family Icteridae, grackles and some blackbirds have eyes light yellow or white, the Bronzed Cowbird has red eyes, while those of most other members of the family are dark. Similar variations exist within the vireo family; in the Red-eyed Vireo, and perhaps other species, the red eye color is found only in adults. In the Surf and White-winged Scoters, males have light eyes and females have dark eyes. Unusual eye colors may play a role in displays; other functions are not yet known.

Hearing

The number and variety of noises that birds make indicate that their hearing is acute. For most birds, hearing is a way of receiving the communications of other birds; few birds use their ears to warn them of approaching dangers, as do so many mammals, and only owls can locate their prey entirely by sound. Some birds, like robins, may listen for the vibrations produced by movements of their prey underground, but this is not yet proved.

Tests of the hearing ability of birds show that they hear within a range that partially overlaps our own. While man's hearing range is nine octaves, Starlings hear about five octaves, from 650 to 15,000 cycles per second, so that every sound lower-pitched than the C note

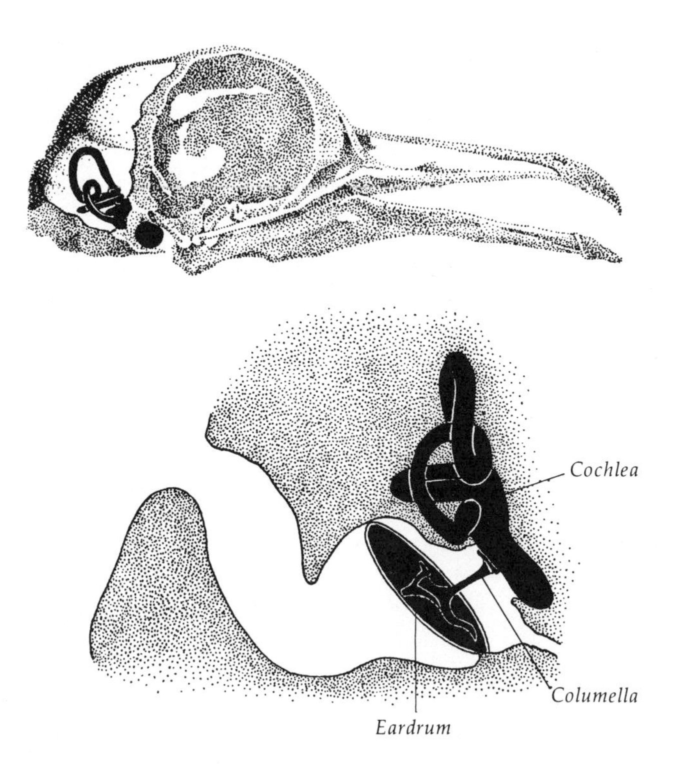

The ear of a common pigeon. Inside the ear cavity, sounds received by the eardrum are transmitted to the cochlea by a rod-like bone, the columella.

at 600 cps (two octaves above middle C on the piano) is imperceptible to them. Most small birds have similar hearing ranges, although warblers probably hear notes higher than we can. If you watch certain high-pitched warblers sing, you may see their throat continuing to quiver and their bill remain open after you can no longer hear any sound; presumably the birds hear all sounds they make themselves, as they respond to recordings of their own vocalizations. (Since most small birds therefore cannot hear human voices, the real reason to be quiet while bird watching is so you can hear the birds.) Some larger birds, including waterfowl, hawks, pigeons, owls, and woodpeckers, which all have lower pitched calls, hear lower notes than do most passerines.

Bird ears are not visible, being covered with loosely constructed feathers called auriculars which do not interfere with sound reception. Even some of the bareheaded vultures have a little group of feathers covering the ear. (The "horns" or tufts on certain owls play no role in hearing.) The ear functions very much as our own, but the length and shape of the cochlea, the fluid-filled inner ear, varies in different species; those with complex songs have long cochleas, but what the relationship may mean is not known. Owls, which have very large ear openings and very long cochleas, also have one opening higher than the other, so that by turning its head to make a sound's intensity equal in both ears, the owl faces it directly.

A few birds active at night have developed a form of echolocation less sophisticated than the bat's ultrasonic system. The birds emit certain high clicking notes and judge the distance and shape of objects from the quality of the sounds bounced back, but can only locate obstacles, not food items. Like many bats, these birds live in dark caves; they include the Oilbird of northern South America, a distant relative of the nighthawks which feeds by picking fruit off of trees as it hovers, and certain swifts of the East Indies.

Smell

In most birds the sense of smell is barely developed, the bones of the nasal passages lacking the sensory nerve endings that mammals, for example, have in abundance. Among the few birds with a demonstrated sense of smell are the tubenoses, which are attracted by oily fish scents from far away. Seabird-watching boat trips often pour small amounts of heated fish oil on the water; shearwaters and petrels far out of view will sometimes appear with astonishing speed.

The Turkey Vulture has long been suspected to use a sense of smell to locate carrion, but experimental evidence has been contradictory. Audubon found that vultures could not detect a strong-smelling carcass if it was covered, but would find and attempt to eat his painting of a dissected sheep. The experiments of others have shown Turkey Vultures able to find hidden animal carcasses but not equally strong-smelling hidden fish carcasses, although the vultures will eat fish when they find it. Recent research indicates that Turkey Vultures can detect at least certain scents, but depend more on sight to find food. The related Black Vulture and condors do not have a sense of smell, nor do the unrelated Old World vultures.

The kiwis of New Zealand, flightless birds that feed at night on earthworms, have poor vision, but they have developed the strongest sense of smell of all birds. Unlike other birds, they have nostrils at the tip of the upper mandible, and seem to sniff the ground for concealed food.

Taste

The sense of taste is often closely associated with smell, but we do not know that birds with a demonstrated sense of smell have greater sensitivity to taste. Birds have very few taste buds, usually between forty and sixty, compared with man's approximately ten thousand, and most of their taste buds are not on the tongue, but on the roof of the mouth and in the throat. Some birds also have taste buds on the edge of the mandibles.

Most birds swallow their food quickly, giving little indication of

The excretory and reproductive systems.

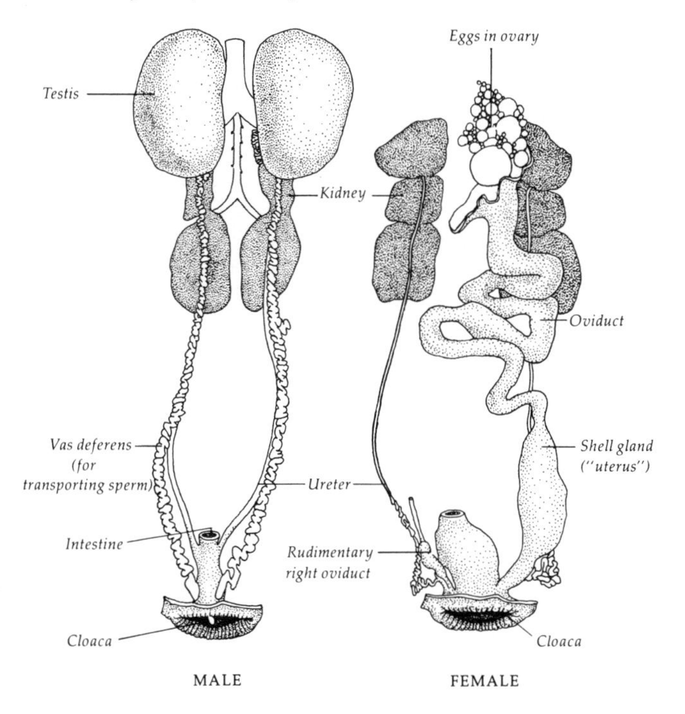

taste discrimination. Insect eaters, however, learn to avoid the Monarch butterfly and a few other insects whose diet of milkweed gives them a bitter taste and some poisonous qualities. At feeders, birds may examine, accept, or reject certain new types of food, but we do not know what role taste has in the decision.

Reproductive Structures

Reproduction begins in birds, as in mammals, when an egg, or ovum, is fertilized in the oviduct by contact with a sperm cell; the fertilized ovum forms the nucleus of the egg, which will be equipped with a food source (the yolk) and a protective shell before laying. Certain domesticated birds like chickens and ducks regularly lay eggs without receiving sperm from the male; wild birds occasionally lay these unfertilized eggs also.

As yet another adaptation for lightness, the male's testes only enlarge when producing sperm; then they become several hundred times the normal size, and often the left testis is larger than the right. Sperm passes to the cloaca in a coiled tube, where it is often temporarily stored; the resulting bulge in the cloaca is used by ornithologists handling birds to judge their breeding condition.

The female has two ovaries; except in raptors, where it varies individually, usually only the left develops, and it too shrinks after producing the season's ova — another probable adaptation for lightness.

Sperm is transferred to the female during copulation by direct contact of the two cloacas. The male briefly stands on the back of the female while the cloacas are pressed together, as shown on page 152. Most birds have no copulatory organ like the penis in mammals, but in the Ostrich and waterfowl an erectile structure on the cloacal wall serves as a penis. In waterfowl this is an important adaptation to assure the transfer of sperm, since they often copulate in the water, but in Ostriches its purpose is not known.

Endocrine Glands

Endocrine glands is the name for the group of glands which secrete chemicals called hormones directly into the bloodstream; hormones have important effects on growth and behavior, often influence other organs in the body, and control essential life processes.

The pituitary gland, located in the skull under the brain, secretes hormones that control the other glands. Each of the two lobes of the pituitary affects different glands. One also directly controls aspects of reproduction and parental behavior such as the instinct to incubate eggs and the development of the brood patch (the unfeathered skin on the underside with which incubating birds keep eggs and young chicks warm). The other pituitary lobe regulates blood pressure, reduces the volume of urine, and may cause (or more often prevent) the laying of eggs before they are fully developed.

The thyroid gland is stimulated by pituitary secretions and controls the growth of the gonads, which produce sperm and ova. Inadequate amounts of thyroid secretion prevent sexual maturity. Too much secretion speeds up molt, and in feathers increases the formation of barbules and dark pigment, while too little has the opposite effects. Body temperature regulation is also affected by the thyroid.

The parathyroids control the amounts of calcium and phosphorus in the blood and influence bone formation. Calcium is also essential to the formation of egg shells.

The adrenal glands serve in conditions of stress to raise the level of sugar in blood, increase heart rate, and raise blood pressure. These are important for sudden bursts of energy.

The gonads produce hormones as well as sex cells. These affect sexual behavior and physical characteristics like the development of brighter plumage in males. Male hormones are present, in different amounts, in birds of both sexes. Females that lose their ovaries, by disease or by operation in experiments, develop male characteristics, occasionally even acquiring the ability to produce sperm. Loss of testes in males, however, does not produce female characteristics, but only an absence of male sexual behavior. Changes in bill color, like the Starling's from dark brown to bright yellow in spring and then back to brown in fall, are caused by gonadal stimulation.

The role of the endocrine glands is still very imperfectly known, especially the factors activating or diminishing hormone production. Some hormones affecting migration, reproduction, or molt seem influenced by changes in day length, while others function independently.

Chapter Seven

Voice

ONE OF the most distinctive characteristics of birds is the variety of sounds they make. Producing these sounds is in many cases a purely inherited and instinctive ability, but recent experiments have shown that some birds learn the sounds they make by listening to other birds around them. The ability to learn vocalizations is limited to birds and man. As in man, the vocalizations of birds communicate many meanings and are essential to survival.

The mechanism of sound production is still not well understood, but, as discussed in greater detail below, delicate muscles in a "voice box" at the lower end of the windpipe control the amount of tension needed to produce sounds of different pitches when air passes vibrating membranes.

Most birds can be identified by their voice; in fact some similar looking species can only be positively told apart in the field by their call or song. By observing birds carefully, you can learn the meanings of many of the sounds they make, so that even without seeing them you will know what they are communicating. To further your knowledge, many good records and tapes are now available with the sounds of birds from all parts of the world.

Recent developments in sound recording and interpreting equipment have opened new fields to the study of bird vocalizations. In the past, bird songs were described in terms of words, sounds, or musical notes, or were distorted on poor records. Now, with audio-spectrograms that represent sound graphically, we can see precisely the sound frequencies of each bird call, the number of notes involved, and their length. Shown in this way, bird sounds can be analyzed in detail. (There are even instruments that can "read" an audio-spectrogram and reproduce the appropriate sounds resembling a bird call.) Audio-spectrograms may be hard to interpret at first,

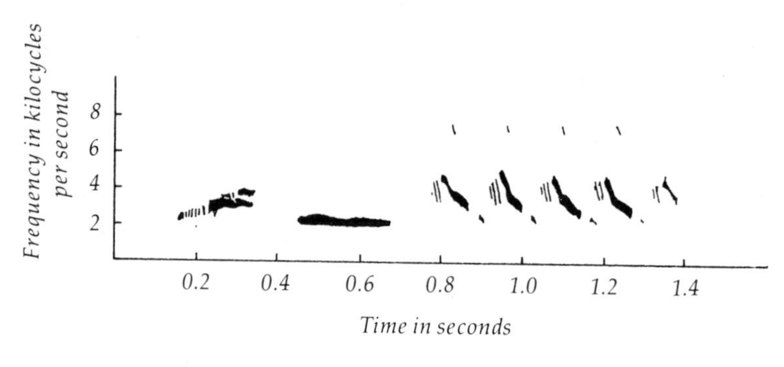

Audio-spectrogram of the Rufous-sided Towhee's familiar "Drink-your-teeee" (from D. J. Borror, CONDOR, *77, 1975.).*

but become easier after you have studied the spectrograms of songs you have actually heard. By starting with familiar ones such as the Rufous-sided Towhee's shown above, you will develop the ability to understand the spectrograms of bird songs you have never heard.

Bird vocalizations have been divided into songs and calls, depending primarily on their message. Those directed at territorial rivals or mates are considered songs; they are usually longer and more complex than calls, which include all the other kinds of messages. Although the passerines are often called "songbirds," some make few sounds that qualify as songs, and many non-passerines have vocalizations whose function and form entitle them to be labelled song. As described below, both calls and songs are strongly influenced by the bird's environment.

Song

As every bird watcher knows, the song of each species usually follows a set pattern, but there is enough flexibility within the pattern for regional and individual variation. Recordings have shown that songs contain many more notes than the human ear ordinarily hears; if you play a bird record at a slow speed you will hear some of these other notes.

Males do most of the singing, since in most species they establish and defend the nesting territory and are first approached by a female looking for a mate. In a few species, including the bluebirds, Northern Oriole, Cardinal, and White-throated Sparrow, both sexes sing, although the males are more active in territory defense and use song to advertise their availability to females at the beginning of the nesting season. In some species, like the Mockingbird, where the sexes have separate non-breeding territories in winter, males and females both sing, but when paired again in the spring, only the male sings to defend the jointly held territory.

Some species have a wide variety of song variations, others give the same notes repeatedly. In either case, any particular song may have more than one meaning. The spring song of a male advertising his ownership of a certain territory warns other males, "this place is occupied and if you cross the boundary you will be driven off." To a female, the same song means, "here is an available male." When the pair bond has been formed between a male and female, the song may stimulate the female to copulation, and later inform her of her mate's whereabouts within the territory. In some species, recognizably different songs communicate each of these messages.

The songs of birds that do not maintain territories serve mainly to advertise the availability of the male for mating, and in species such as the American Woodcock and hummingbirds where the male has no role in nesting or raising of the young, the song and associated displays are only to attract a female for copulation.

Although most birds use their voices to communicate, there are other ways. Many woodpeckers hammer on hollow branches or other objects to make a loud noise. Air rushing between the feathers as a Ruffed Grouse stands rapidly beating its wings produces a drumming sound; this noise announces his availability to females. Grouse of the plains, like the prairie chickens and Sage Grouse, gather on courtship grounds where the males produce a variety of loud sounds by rattling their feathers, stamping on the ground, and passing air through inflatable air sacs near the throat. Such a scene is shown on page 142. Hummingbirds, snipe, and nighthawks, among others, produce loud screeching noises when air passes through wing or tail feathers in the dives of courtship flights.

Annual Song Cycle

Since songs communicate specific messages, they are usually delivered only when the message is appropriate. Birds defend a territory and try to attract a mate at the beginning of the nesting season; this is when you will hear the most vigorous and continuous singing. For most birds, these activities take place in spring, but species on different schedules sing at whatever time of year it serves their purposes. Great Horned Owls, which begin nesting in late January and February, are very vocal then. The European Robin, a thrush not closely related to the American Robin, sings frequently in December, when it seeks a mate, although the pair does not begin nesting until late March or April. Species like the Mockingbird which use song to communicate ownership of a winter territory sing at that season too.

Song, like other behavior patterns related to reproduction, is stimulated by the development of the gonads. For many birds, the annual release of the hormones that develop the gonads is thought to be stimulated by increases in day length. Thus, for birds away from the Equator, where day length is the same the year round, the beginning of song seems related to the increase in day length which comes every spring. The stimulation is evidently so strong that many birds sing before it serves any function directly related to reproduction. On bright winter days White-throated Sparrows still on their wintering grounds often sing. Many birds sing during their spring migration, when they are not yet defending a territory or seeking a mate, but these songs, like those of winter singers, may not be complete or exactly like nesting-area songs. Why some species sing more than others on migration is not known. Within a closely related group of thrushes, the Veery and Swainson's Thrush sing commonly during spring migration, the Gray-cheeked Thrush sings infrequently, and the Hermit Thrush rarely if ever. One possible function of late winter and migration singing may be the "practice" it gives birds before they reach the place where correct singing is crucial; "practice" might be especially valuable for birds about to breed for the first time.

Having performed its important functions — communicating territorial ownership, desire for a mate, and maintenance of a bond between a pair raising young together — song gradually tapers off. By midsummer, when many young have left the nest and territories are breaking up, few birds are singing except those that have lost a mate or are renesting. Different species cease singing at different stages of the nesting cycle; those in which the male sings only to attract a mate, but does not defend a territory or help the female raise the young, sing for only a brief period. The American Redstart sings constantly until nest construction is completed, then ceases. The Common Yellowthroat, in contrast, sings during every stage of the nesting cycle except courtship. The House Wren continues to sing while raising the young; it even sings while carrying food to the nest.

The shrinkage of the gonads after having produced sperm and eggs may also affect the reduction of singing during the summer. For a reason not known, the gonads of many birds briefly enlarge and then shrink again in fall. During the period of fall gonad enlargement birds may sing again; the songs at this period are often softer and less complex than spring songs. Gray Catbirds, American Robins, and White-throated Sparrows are among the birds that can frequently be heard singing in September and October.

Daily Song Cycle

The majority of birds sing most frequently and energetically early in the morning, often beginning before dawn. Birds usually leave their sleeping place before singing; many sing each morning before looking for food. The time when each species begins can usually be correlated with the amount of daylight; certain species regularly start earlier than others. Insect eaters often rise and begin singing before seed eaters, because they can find food more easily in dim light than seed eaters can. Hole-roosters are often later risers, since the first glimmer of light does not reach them.

After the early morning burst of song, birds sing less frequently, often stopping entirely in early afternoon when many birds seem to rest. Late afternoon and dusk is a period of renewed song activity, although rarely as intense as early morning. A few birds sing all day. The Red-eyed Vireo is especially well known as a persistent singer — one closely observed individual gave its one-second song more than 22,000 times in ten hours. Some birds continue to sing at night; the Mockingbird and Yellow-breasted Chat often sing throughout the night, especially when there is moonlight.

The American Woodcock performs its song flights during darkness, beginning regularly a half hour after sunset and continuing until total darkness. If there is a bright moon, the woodcock may continue off and on through the night, otherwise it begins again when light intensity reaches a certain low level before dawn. Other night birds, like owls, Whip-poor-wills, and rails, have regular song periods dependent on the amount of light. The Chuck-will's-widow sings most with moonlight; the Black Rail prefers total darkness and is most vocal on nights without a moon.

Some birds give different songs at different times of the day. Thrushes, the Willow Flycatcher, and the Ovenbird, which usually sing from a perch or the ground, have evening flight songs. The flight songs may be restricted to the time when they make the bird least vulnerable to predators. At twilight the Eastern Wood Pewee sings a faster and more complex song than the familiar whistle given all through the day. Other birds have songs given only before dawn. Little work has been done on whether the less frequently heard songs have meanings different from those the bird sings all day.

Song and a Bird's Environment

Animals communicate by whatever means can be most easily understood: many mammals indicate the limits of their territory with scent marks; moths use scent to attract a mate; brightly colored tropical fishes use their bodies as flags to visually communicate territorial ownership. For most birds, sound is the most efficient medium of communication. The song of a small bird, especially in a habitat of trees, shrubs, or tall grass, can be heard much farther away than the bird can be seen. In addition to traveling farther than any visual communication, song does not require a bird to expose itself conspicuously to predators as visual communications do. Birds like prairie grouse that use visual displays in addition to songs and calls display in open places where a predator could not approach unseen.

Like anatomy and plumage, songs and calls have evolved to reflect the demands of each species' environment, and as with anatomical and plumage characteristics, there is often convergence in the vocalizations and associated behavior of unrelated birds that inhabit similar environments in different parts of the world.

The first requirement of a song is that it be heard. Songs communicating territorial ownership or availability are loud. Songs directed at a mate within the territory need not be as loud, and songs associated with copulation, courtship feeding, or other activities where the pair is already together may be very soft. Similarly, the fall and spring migration songs, which do not communicate anything to other birds, may be soft.

The variety in bird songs reflects the fact that different sounds travel better in different environments. A song to be heard in neighboring territories must be not only loud, but of a quality that travels easily. The waves of high-pitched sounds are shorter than those of low-pitched sounds and more easily blocked by solid objects, so a sound that must travel through an area of many obstacles will travel farther if low in pitch — the songs of the warblers illustrate this principle. Species singing from the ground or very close to it, like the Ovenbird, waterthrushes, and Kentucky Warbler, need songs that will travel past tree trunks and other large obstacles at ground level. Their songs are the lowest in pitch of any in the wood warbler family. (Waterthrush songs, which must also compete with the sound of running water, are especially loud.) Species singing from shrubbery or low trees have somewhat higher-pitched songs; the Common Yellowthroat and the Wilson's and Yellow Warblers are in this category. All of the very high-pitched warbler songs are given by species like the Blackpoll, Bay-breasted, and Blackburnian Warblers, which sing from the tops of tall trees, where their songs encounter few obstacles.

A bird that uses song to make its exact location known so as to attract a mate or inform the mate of its whereabouts must produce sounds that can be easily located. Pure tones, like a high or thin whistle, are much harder to locate than buzzing, rattling, or repetitive sounds. Birds that sing while concealed in marshes, for example, where there is very little visibility, have songs that are repetitive, like the Swamp Sparrow's, or rattling, like the Long-billed Marsh Wren's. Red-winged Blackbirds, singing in the same habitat, but from a more elevated perch, have a hoarse song, also easily located.

Birds inhabiting open, flat country — meadows, prairies, northern tundra — often prefer to sing from some vantage point above the ground; this enables them to better survery their territory and makes neighboring birds more aware of their presence. Their songs also travel farther when delivered from high over the ground. Meadowlarks and grassland sparrows generally sing from the few trees, bushes, wires, or fenceposts in their territory. Individuals have regular song perches within their territory, and may even choose a territory on the basis of suitable singing places, rejecting areas without any.

Birds of uninterrupted prairies or treeless tundra where there are no elevated perches sing while in flight, often from very high up. The Skylark of Europe (introduced in British Columbia) gets its name and its famous reputation from the long, beautiful song it delivers while hovering over its territory. In North America, the Bobolink, Lark Bunting, most longspurs, and several sandpipers sing primarily from the air, and, as noted, some birds of wooded habitats occasionally do so. Song flight is different from a bird's other flying styles; while singing, birds usually hover on quivering wings or fly in circles. Some, like the Skylark, continue singing as they slowly glide to the ground. Most birds that regularly sing from the air have contrasting white marks on the tail — highly visible when it is spread in hovering flight; these patterns make the bird all the more noticeable to its neighbors.

While certain ecological factors make songs of certain sound types most effective in certain habitats, other factors prevent the songs of species living together from becoming too similar. Songs are communications intended only for members of the same species; if birds mistook the songs of other species for their own, they would lose much time and energy in useless encounters. In environments containing many species, the song of each is richer but less varied than where they occur with fewer neighbors of other species. This is demonstrated by the rougher or more variable songs often heard from island populations of widespread birds — on the island there is less reason to develop distinctive songs, but on mainlands the pressure of many species living together forces each to restrict its song to a highly recognizable pattern.

Geography may further affect bird song in that widespread species sometimes have different songs in different parts of their range. This is especially true when populations are isolated from one another. Song, like all other characteristics controlled by the genes, is always subject to evolution. The Screech Owl and White-crowned Sparrow are two species with songs so different at opposite ends of their ranges that eastern and western individuals do not even respond to recordings of the others' songs. Since song is one of the ways individuals recognize other members of their own species, when this mechanism fails it indicates that separated populations may be evolving the behavioral barriers which make reproduction impossible and lead to distinct species. Of course, much more than differences in song type is needed to prevent reproduction between closely related populations. The very closely related Eastern and Western Meadowlarks have completely different songs, whose main function seems to be territorial communication between males, since females recognize males of their own species by their call notes, not their song.

To further avoid wasting energy in singing or territorial fighting, it is useful for birds to recognize the songs of individuals of their own species. Most can recognize the songs of their mate and of their neighbors; White-throated Sparrows, for example, can distinguish between recordings of their neighbors and those of strange birds, and will react more aggressively to the song of a White-throated Sparrow they have not heard before.

Weather and climatic conditions may also exert evolutionary pressure on bird song. One example is the "duets" sung by mated pairs of many tropical birds, including various quail, rails, owls, wrens, and shrikes. In some species the pair sings the same or different notes at the same time; in other species the male sings a few notes and the female immediately replies with the others — sometimes so quickly that recording equipment is needed to show that two birds are involved. In all cases the duet songs are highly stereotyped and unvarying. The function of these songs is to keep a mated pair together — both within the territory and "psychologically," so that when the breeding season begins no time will be lost searching for a mate. Two types of ecological pressure make this a useful adaptation. In tropical forests many birds are uncommon and widely scattered; a mate is not always easy to find. For birds whose nesting season starts with the irregular rainy seasons of some tropical and desert areas, maintaining the pair bond with duetting assures that birds will be ready to breed as soon as the sudden rains make environmental conditions right. In North America, where the nesting season begins more predictably, no birds use duets to maintain a pair bond, but members of a mated pair often call to one another. Duetting probably evolved from these calls rather than from territorial or mate-attracting song, but duets are now so complex and musical that they are usually considered a kind of song.

Calls

Even during the season when birds sing, the songs do not communicate all of a bird's concerns. Aggression, alarm, danger, and food location are other kinds of information some birds convey by the short, unmusical notes labelled "calls," often heard throughout the year. By watching a bird's behavior as it makes different calls, you can understand their meanings.

The number and types of calls made by birds varies. Vultures have a few hissing or muffled barking notes; no sounds are reported for the California Condor, which conveys courtship, alarm, and disputes over food in visual displays rather than by sound. Other silent birds use nonvocal sounds — storks greet one another with repeated claps of the mandibles. Even birds living in groups, like certain pigeons that have many interactions between members of a flock, may be relatively silent, communicating by gestures and displays.

The variety of vocalizations has been especially well studied in gulls and some passerine groups. Probably the greatest number of calls are given by members of the crow and jay family — perhaps another indication of their purported high level of brain development. Observations of these especially vocal birds have shown that many calls are accompanied by certain body movements to make rituals so specific that they will not be confused. It is important that all members of the species understand the meaning of each call — to misinterpret a call could be dangerous.

Aggressive calls are among the most widespread in birds. Dis-

This Herring Gull is in the posture associated with the final stage of what behaviorists have named the "trumpeting call." It is a challenge that precedes — or prevents — aggressive encounters with other individuals.

putes may arise over possession of territory, a mate, nesting material, or food. Calls are a signal of how aroused, belligerent, or timid each bird is, and combined with specific gestures, they limit most such encounters to bluff and display, minimizing the amount of actual fighting and physical damage; one bird usually retreats before an actual fight begins.

Calls informing others of a bird's location are also common. These calls help members of a pair or a flock stay close or, in other cases, keep far enough apart. Calling between members of a pair is particularly common among birds living in environments with little visibility. In some species, the male may call or even give a special song informing the female on the nest of his location. Among goldfinches, where there is individual variety in call notes of the same meaning, members of a mated pair develop increasingly similar notes, making recognition easier. Birds flying in flocks, like Cedar Waxwings, often call continuously; these calls may maintain a sense of unity within the flock. In contrast, the chirps often heard from birds migrating at night may prevent collisions between birds that cannot see one another. Calls of birds feeding in loose flocks may inform others to keep their distance so that individuals will not compete for food items.

Calls conveying the location of food are made by nonterritorial birds like gulls when they find more than they can consume themselves. Similarly, the calls given by a flock of waxwings or finches while feeding may inform others passing by of a food source. Some territorial birds have a call giving the mate food information.

During courtship and the nesting cycle many birds also have calls used between a pair related to copulation, nest site and construction, feeding of the female while she incubates, changeover at the nest if both sexes incubate, location of the young, and other important concerns.

Danger or alarm is a frequent communication among birds, often recognized and understood by other species; calls of birds suddenly flushed from the ground or trees may convey their alarm to others. Geese, crows, and a few other birds feeding in flocks have at least one member of the group watching for danger while the others feed; its alarm calls notify the entire group. The easily recognized, sharp alarm notes of Blue Jays have a sudden effect on other birds, making them silent and still. Birds have alarm calls of different intensities, depending on the danger. Many also have an "all's well" call, used after the danger has passed.

Many calls are interchanged between parents and their young. Ducklings, among others, make cheeping noises for a few days *before* they hatch from the egg, while the incubating female calls to them. This interchange may help the young to recognize their mother when they hatch. Chickens and other birds whose young leave the nest and feed themselves immediately after hatching have calls indicating the presence of food. Many parent birds have calls giving the young their location, telling them of danger, and noting when the danger is over. Chicks fed by their parents beg noisily for food. Usually, the noisiest or most vigorous beggar is fed most, so this is an important vocalization. Young birds give different notes just after they have been fed, and when they are full, contented, and warm. They also have distress calls, given when cold, wet, or confronted with an enemy.

The character of calls, as with songs, is affected by the environment. When small birds spot a passing predator, particularly a hawk, their alarm calls are high and thin, because such notes are much harder for a predator to locate than either low, modulated, or buzzy calls; some of the high-pitched alarm calls may be above the hearing range of hawks. Many of these same small birds will gather in a mixed group and "mob" or harass a perched hawk or owl, approaching closely, scolding, and trying to drive it away. The scolding notes are very different from the alarm calls given when the bird did not wish to be found; they are harsh and easy to locate, enabling other birds who understand their meaning to quickly join the mob. The similarity of such calls in small birds from many parts of the world suggests convergent evolution due to the same environmental pressures.

The calls of young birds reflect similar pressures: begging calls, especially of birds which have left the nest, are loud and rough, enabling parents to find them easily; these begging calls are often higher than any made by adults, and may be above the hearing range of predators. Distress or alarm calls are predictably thinner and more difficult to locate. As young birds mature and become independent, their voices acquire the characteristics of adults, although they will lack the full repertoire of songs and calls until they actually need them, when breeding for the first time. Some of the juvenile calls serve a purpose in adult life; when given food by the male during courtship, a female often assumes the posture and gives the calls of a begging young bird.

Mimicry

Vocal mimicry, the ability to imitate sounds one has heard, is limited to birds and humans. It is one of the indications that sound production is not just an inherited, purely instinctive ability; birds, like people, are capable of learning new ways of operating the mechanisms that produce sounds.

Mimicry is found in several passerine families, in parrots, and in cuckoos. In most cases its function is still unknown. Mimics are found in habitats from bare open ground to dense scrub to forest; some mimic while in flight, others from the ground or a concealed or exposed perch. This variety makes it difficult to suggest ecological reasons why the ability to mimic is advantageous.

North America's ablest and best known mimic is the Mockingbird. Other members of its family, the Brown Thrasher and Gray Catbird, for example, are also good mimics, as are crows, jays, and Starlings. All are able to imitate other bird calls and songs, animal sounds, and mechanical sounds. Some can reproduce a sound as soon as they hear it, others must hear a sound several times. Most mimics are stimulated by the sounds they hear. In spring a Mockingbird will reproduce the songs of returning migrants as soon as they arrive; at the end of the spring it sings many more songs than early in the season. The Mockingbird may continue to sing songs of birds that have departed or ceased singing themselves. Captive Mockingbirds may reproduce sounds years after they last heard them. Like other mimics, the Mockingbird is sometimes stimulated to mimic a particular sound when it sees the bird or object which makes it. While the Mockingbird seems to reproduce other birds' songs very accurately, the other species are not fooled, perhaps because the Mockingbird sings so many different songs in succession; it would be an inconvenience, not an advantage, to any mimic if it actually fooled other birds.

"Talking" birds — those that can imitate human speech — are less common than mimics. The Gray Parrot of Africa is considered the best talker, but many other parrots, as well as mynahs, magpies, and crows, are able talkers. The European Robin and the European Blackbird, a thrush in the same genus as the American Robin, have occasionally learned to mimic human speech. It is curious that in the wild the vocabulary of most parrots is limited to raucous shrieks; that they imitate no sounds in the wild seems to indicate that the ability has no adaptive significance.

Talking birds do not really understand the meaning of what they say and cannot put together sentences they have not heard spoken, even if they have used in other phrases all the words needed for that sentence. Some birds, parrots especially, have remarkable powers of association and will say "good-bye" if they see someone leaving or, occasionally, someone they don't like arriving. Parrots may also repeat a phrase only when the appropriate person is present. One Gray Parrot could whistle music by Beethoven when only the name of the piece was mentioned.

Vocal mimicry has evolved twice to serve an important biological function. In Africa and India, species of cuckoos and indigobirds (in the family of the House Sparrow) that lay their eggs in the nests of other birds, which will raise the young as their own, have calls that mimic their host species. The young of a cuckoo called the Koel has begging notes like those of its usual nestmates — Indian Crows. The young Koel evidently learns to imitate calls by hearing them; Koels raised without Indian Crows do not use their call. Indigobirds include the song of their usual host in their own; the imitative song of the male indigobird may draw the host female away from the nest so that the female indigobird can deposit her egg. As each male mimics the song of only one of the species' several hosts, this may help female indigobirds recognize appropriate mates, since each lays eggs in the nests of only one species.

How Vocalizations Develop

Vocalizations are an important part of the life of most birds from the time they hatch. As noted, some birds begin calling even before hatching when they pierce the membrane of the inner shell, giving them access to the egg's air space. Even species that are practically silent as adults make begging noises as chicks.

Recent experiments have revealed the variety of ways birds develop their vocal repertoire. In some species calls or songs are entirely inherited, in others they are learned by listening to other birds. To test whether birds need to hear the calls of their parents or other birds in order to develop proper vocalizations, researchers have deafened chicks immediately after hatching. Chickens, turkeys, and pigeons deafened at hatching developed perfectly normal calls as they matured, indicating that their vocalizations are entirely inherited. Further proof came from rearing undeafened chickens, turkeys, and pigeons with foster parents of other species — these birds were entirely uninfluenced by the sound they heard around them, producing only their own calls.

In passerines the development process is more complex. Deafening Canaries significantly, but not entirely, alters their song; deafened Song Sparrows do not sing their own song. Eastern and Western Meadowlarks, Cardinals, and White-throated Sparrows reared in complete silence developed poor songs, which were, however, better than those produced by deafened birds. Hearing itself evidently improves a bird's singing ability. Still more influential for these birds that do not inherit all their vocal "knowledge" is hearing song around them while growing up. The effect of other bird sounds in the environment varies widely. When birds hear the sounds made by their own species, they are not influenced by any others, but when raised in unnatural circumstances the effects are different and revealing: European Robins raised by Nightingales, another thrush, will later sing a Nightingale's song; young White-crowned Sparrows exposed only to the song dialect of a different white-crown population will sing the dialect they have heard; Song Sparrows raised by Canaries still sing a good, but not perfect, Song Sparrow song.

Most song learning is accomplished in the first year of life, before young birds breed for the first time. Once the vocalizations are acquired, they are rarely lost; birds deafened after their first year hardly alter their song. Whether raised hearing their own song or influenced by another race or species, Cardinals and White-crowned Sparrows do not alter their song pattern after the first year. Few birds other than mimics develop new calls later in life. The call notes of Cardinals, unlike their songs, are unaffected by deafening, silence, or the influence of foster parents, and so seem entirely inherited.

In the wild, song development begins early for many passerines — while still in the nest, some begin warbling a jumbled series of notes containing elements of mature song. This kind of "practice" singing, called subsong, becomes increasingly like adult song; it has no communicatory function, and is usually given by chicks and fledglings while relaxed, at a low volume. Some Song Sparrows begin warbling at 13 days, American Robins at about 21, and Brown Thrashers at 44 days old. Subsong stops in winter and is resumed in early spring; by the time an accurate song is needed for territorial or courtship functions, the bird has perfected its technique. The singing of mature birds for a brief period in fall and during spring migration benefits first-time singers who need to "tune up" and can imitate those with experience.

Subsong occurs only in species that learn their song from what they hear around them. In some cases, young learn their song pattern directly from their father: a European Bullfinch raised with other Bullfinches by a Canary sang a canary-like subsong and mature song, and its own young, although they heard other Bullfinches singing normal song, learned and repeated their father's canary-song. In a carefully studied group of wild Song Sparrows, it was found that no individual sang exactly like its father or like either grandfather; each bird had individual variations of the general Song Sparrow pattern.

How Sounds are Produced

Vocal sounds are produced by the muscles of the syrinx, or voice box, which is located at the lower end of the windpipe. A few birds, like the Turkey Vulture, have no syrinx and make only an occasional grunt or hiss. Storks and ostriches have a syrinx, but not the muscles necessary for sound production; they are limited to hisses, grunts, and booming noises.

The position of the syrinx varies in different groups. The syrinx of some passerine families limited to Central and South America is an expansion at the lower end of the trachea. Cuckoos, nighthawks, and some owls each have two syringes, each located in the upper end of the bronchi, the two passages between the windpipe and the lungs. The majority of birds have the syrinx at the bottom of the windpipe, where it divides into the two bronchi. The respective advantages of these systems are not known — there are able vocalists among the birds with each arrangement.

The drawing on page 128 shows the parts of a typical syrinx located where the trachea joins the bronchi. The tracheal and bronchial rings are expanded to form the tympanum, where sound is produced. The semilunar membrane, at the tip of the pessulus jutting into the tympanum, vibrates when air passes out of the lungs and up

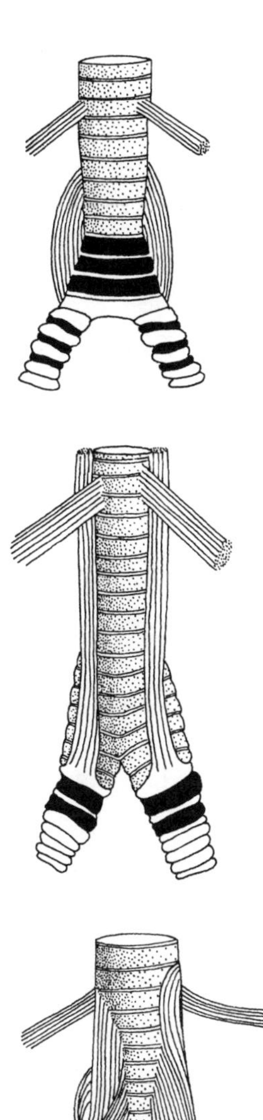

The three types of syrinx found in birds: from top to bottom, the tracheal syrinx of a few New World tropical families including the woodcreepers and antbirds; the bronchial syrinx of oilbirds, nighthawks, and some cuckoos; and the tracheobronchial syrinx of most other birds.

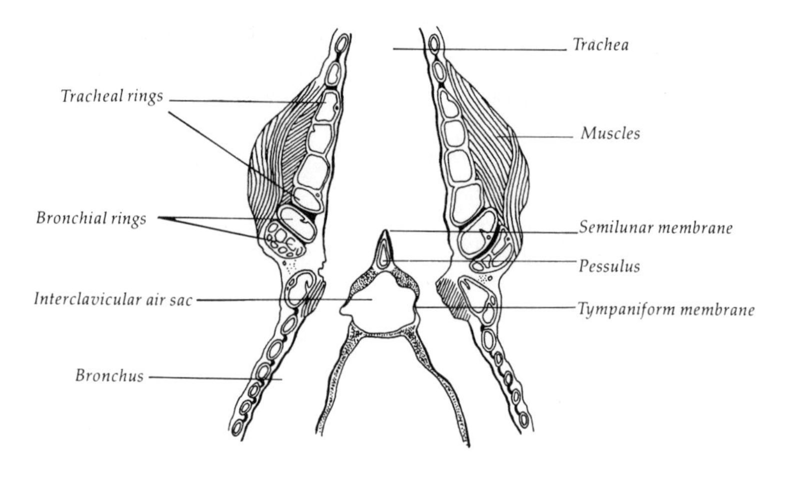

The trachea of a European Blackbird, a close relative of the American Robin (from BIRD SONG: ACOUSTICS AND PHYSIOLOGY, *by Crawford H. Greenewalt, Smithsonian Institution Press.).*

the windpipe. Underneath the pessulus the tympaniform membranes form the narrow openings out of the bronchi; these membranes are controlled in different species by from one to eight pairs of muscles. Air coming out of the lungs presses against the tympaniform membranes, making them vibrate. With delicate adjustments the muscles control the frequency of membrane vibrations, producing sounds of different pitches. While on a piano or guitar different pitches are produced by vibrating strings of different length, the bird uses different degrees of tension produced by the syringeal muscles to achieve variety. The function of the semilunar membrane's vibrations is not clear, since when it was removed in experiments with Starlings there was no apparent change in vocalizations.

The syringeal muscles are a good indication of vocal ability: they vary from one to eight pairs, the highest number being found in mimics and in singers that produce a wide variety of sounds, like the Common Crow, Mockingbird, and Starling. The arrangement of the muscles is very regular within families and is useful in taxonomy. The many-familied passerine order can be divided into "oscines" and "suboscines" — terms frequently used — by the arrangement of these muscles. Oscines have the syringeal muscles attached to the ends of the bronchial rings, which enclose and protect the syrinx; all North American passerines except the flycatchers are oscines. In suboscines the syringeal muscles are attached to the middle of the bronchial rings. Many of the suboscines have whistle and flutelike notes, but the most elaborate songs are all produced by oscines.

Chapter Eight

The White Tern builds no nest, laying its single egg on a horizontal branch. It breeds on subtropical oceanic islands in the South Atlantic and tropical and subtropical islands in the Pacific.

The Breeding Cycle: Territory, Courtship, Nest, and Eggs

THE ADAPTIVE significance of many of the physical and behavioral characters we have examined is made clear by their role in the breeding cycle. Birds have evolved many ways of mating and rearing their young; these are often called "reproductive strategies" because, within each species, they follow a fairly fixed plan which, repeatedly tested by natural selection, has become the most efficient manner of reproduction for that species in its environment. (They are not strategies in the sense that a bird stops to calculate the alternatives.) All reproductive strategies that have been studied share the basic principle that birds produce, on the average, as many young as can be successfully raised. Birds do not limit the number of eggs they lay or young they raise in order to prevent overpopulation — natural predators or food shortages will later bring the population down to whatever number the environment can support.

You may note that all bird reproductive strategies share certain features: a male and female must come together to fertilize the eggs within the female, the eggs must be laid in a safe place where they will be warmed so that the embryo can mature and hatch, and the young must receive enough food to grow until they become independent. The enormous variety of ways birds meet these universal demands, as well as the requirements of their particular environment, make the study of reproduction especially fascinating. Many excellent books have been written on the life histories and reproductive behavior of individual species or groups of birds (some of these are listed in the Further Reading section, beginning on page 281) and the scientific magazines always have articles on details of the breeding cycle, but much is still unknown about some of our most familiar birds. The information in this chapter will give you an idea of what to observe in the reproductive behavior of birds around you and how this behavior is adapted to meet the challenges of the bird's environment.

Beginning of the Breeding Cycle

The breeding cycles of birds are timed to take place at the season when the maximum amount of suitable food is available for the young, or when the most food is available to the female for forming eggs inside her. Sometimes the breeding season represents a compromise between these two important demands. In the Northern Hemisphere, these demands are most often met in the spring, but crossbills will nest any time between January and August that they find an adequate cone crop, and the American Goldfinch breeds in late summer, when thistle seeds are available to feed the young. Eleanora's Falcon, which nests around the Mediterranean basin, begins nesting in July and August to feed its young on southbound passerine migrants in August and September; it is the latest nester in Europe. In tropical parts of the world, breeding is often tied to regularly alternating wet and dry seasons. In the Australian deserts where rains are completely irregular, some birds begin breeding as soon as rain falls, whatever time of year.

Most North American birds do not winter where they breed; in spring, migrants often return to the region where they were born or to territories they occupied before. Males usually arrive first, and begin the types of behavior such as singing and display that will assure them of getting a mate when the females arrive. With some early migrants like the Red-winged Blackbird there may be a period of several weeks between the arrivals of males and females. On spring migration, male warblers are often observed for a week or ten days before females begin to pass through an area. Most of the species in which the males arrive first are ones that defend a territory where mating and nesting take place. The female may not arrive until there is enough food available for her to form eggs; by then, territories are already set up and the male can devote more attention to her.

Certain environmental conditions may make the arrival together of males and females a more efficient system: ducks and geese that nest in the Arctic, where the breeding season is very short, arrive together already paired and lose none of the little time available in preliminaries like courtship. Ducks from many areas winter together, but pairs always return to the area of the female's birth, perhaps because it is more important for the female to be familiar with the area in which she will nest and singlehandedly raise the young.

The age when birds first begin breeding is also subject to environmental pressure. Most smaller birds first breed in the spring following the year of their birth, when they are about ten months old. In some passerines like the Red-winged Blackbird and Starling, females begin breeding in their first year, but males not until their second; perhaps the older males are better able to control all the suitable nest sites, since in experiments where adult Red-winged Blackbird males were removed from a marsh, first-year males, free of competition, were able to breed.

Among larger birds whose feeding techniques require skill and practice, many wait a few years before breeding. (You can distinguish birds too young to breed because their plumage is usually somewhat different from that of adults.) Common Terns do not nest until the third year after their birth, and Herring Gulls not until the fourth. Bald and Golden Eagles first nest when three to five years old. Wandering Albatrosses never breed younger than nine years old, and often not until they are eleven. In all these species the individuals that wait longer, gaining more personal experience before breeding, end up producing more young than do individuals beginning sooner. Among the colonial species like the Common Terns, young nonbreeders will spend the summers at the colony of their birth, gaining experience that will be useful to them when they do breed. When these birds finally nest for the first time, they are still usually less successful than older individuals that have nested before.

At the other extreme are birds whose environment makes the ability to nest when very young more valuable. In Australian deserts where more than a year may go by without the rain necessary to create breeding conditions, the bird populations must reproduce as quickly as possible during the few months when conditions are favorable; some of the small finches and the Budgerigar (the common pet parakeet) resident there can breed when only two months old.

Territory

During the nesting season nearly all birds have some space that they defend against other members of their species, excepting their mate and young. In colonial nesting birds the space may be no larger than the area a bird can reach with its bill when it sits on the nest, while large raptors may defend several square miles. The size and functions of this territory vary with the ecological requirements of each species. Territories serve the following main functions:

1. Spacing of pairs, providing each with a place to mate.
2. Spacing out of nests, making eggs and young harder to find by predators.
3. Even distribution over the available habitat so that enough food can be gathered to feed the young in each nest without unnecessary competition.

For birds nesting in dense colonies, the territory serves solely to give the occupants of each nest enough room to move around, since even mating may take place away from the nest site. However, as discussed below, the ecology of colonial birds is such that the apparent disadvantages of nesting in groups are far outweighed by other benefits.

The vast majority of birds hold territories which serve all the functions listed above. Their food is evenly distributed over areas of suitable habitat and they defend an area large enough to supply the food needs of the parents and young. By defending an area producing enough food, the birds are assured an area with privacy for mating and a nest site away from those of neighbors of the same species. For some birds, hole-nesters especially, locating a nest site is harder than finding a territory with feeding requirements; chickadees, for example, will defend much smaller territories when birdhouses increase the number of nest sites available. Other birds have separate feeding and nesting territories: the Belted Kingfisher nests in a tunnel dug into a bank often far from the stream or pond it defends as a feeding area. Arctic plovers and sandpipers defend a territory around the nest, but as soon as the chicks hatch, the entire family leaves it for another, undefended feeding area.

In some species, the male establishes a territory exclusively for mating purposes. He drives other males away and attracts one or more females with his song or display. After mating, the female establishes her own territory, usually more quietly and with less competition from other females, in which she lays her eggs and raises the young without help from the male. The Ruffed Grouse, American Woodcock, and hummingbirds are among those with separate territories for each sex.

Territorial ownership is most frequently conveyed by songs and displays. The Red-winged Blackbird, for example, sings from a perch over its marshy territory, while it spreads its tail, lowers its wings, and raises its red shoulder patches. Experiments have shown that the red shoulder patches impress rival males more than the song does; males with shoulder patches painted black were less successful in keeping territories or attracting mates. Most displaying birds emphasize whatever is most colorful or distinct in their appearance. The male Ruby-crowned Kinglet raises its crest, the male American Redstart spreads its tail, and the male Ruddy Duck shows its bright blue bill.

By communication in songs and displays, most fights are avoided. Occasionally a trespassing neighbor or stranger is not put off, and an actual encounter takes place, with the two birds facing one another singing or displaying. If the trespasser still remains, a fight or chase may develop. The bird on its own territory always has the psychological advantage and, unless remarkably weak, drives the other away; if he chases the rival back onto the rival's territory, he loses the psychological advantage, and the bird now on its own territory can easily drive the original chaser out. In the wild, such fights rarely harm either bird, because one can always leave the territory in question. In cages, where there is no escape or hiding place, otherwise peaceable birds like doves may fight until one is killed. Very few birds normally chase individuals of other species out of their territory, unless they are possible predators; hummingbirds and Eastern Kingbirds are particularly well known for chasing larger birds away, while Robins, Blue Jays and Common Grackles frequently chase squirrels, which sometimes eat bird eggs.

The size of a territory depends on the ecology of the species and the quality of the environment. Where food is abundant, birds defend smaller territories; where it is sparse they need larger areas. Most small passerines use about an acre. When food is especially dense, as in spruce budworm outbreaks in evergreen forests, warblers may use less than a half acre. Raptors need much larger territories — Golden Eagle territories as large as 75 square miles have been found. Bald Eagles, feeding primarily on fish, which are more abundant than the Golden Eagles' mammal prey, have smaller territories, depending on the amount of fishable water in the area.

Territory shapes usually follow the contour of the land; an approximately circular or oval area is most easily defended. Birds whose territories are based on special features, like a stream for a Belted Kingfisher or Louisiana Waterthrush, or a lake for a Common Loon, do not defend any of the surrounding area, which is of no use to them.

The main ecological feature shared by all birds that defend a feeding territory is that their food is evenly distributed throughout the defended area. Feeding competition from other pairs would make it impossible for them to find enough for their young.

Types of food not evenly distributed, but found in large concentrations, or at changing localities, require another reproductive strategy. Carrion feeders, for example, cannot regularly expect to find food in any given area; defending a territory would be a waste of energy, since the bird is physically incapable of patrolling an area large enough to assure the presence of carrion at all times. Further, much carrion comes in sizes too large to be consumed entirely by only one individual — or one family — and the bird could not defend it. So for birds like vultures and condors, it is much more efficient to share the search areas and the food itself, which is found most efficiently when groups of vultures are searching.

Similarly, seabirds feeding on moving schools of fish, swallows and swifts feeding on moving clouds of flying insects, or tropical birds feeding on fruits or seeds that are ripe in different places at different times, are all better off without separate territories. There are in fact positive advantages to group feeding for these birds, since their food, wherever it occurs, is abundant, and all are guaranteed a share if they follow others that have already located it. If food is concentrated at a few areas, even changeable ones, it is not efficient for birds to scatter widely to keep their nests apart — too much time would be lost in traveling between the feeding area and the distant nest site. Thus it is simpler for these birds to nest close together in colonies near the feeding areas. If you survey all the birds that breed near you, you will find that there is a definite relationship between food distribution and nest placement; birds whose food must be found item by item are evenly scattered on separate territories, while birds whose food is in large quantities at variable locations will nest in colonies.

The disadvantage of colonial nesting is that it makes the nests easy for predators to find, but colonial birds can usually make up for this by nesting in places that are hard to reach. Some swallows and swifts choose bluffs and buildings, herons nest in marshes, and seabirds often nest on cliffs, isolated beaches, or islands. The colonial tropical orioles called caciques usually place their nests in trees with active wasp or biting ant colonies. Colonial seabirds, with habitats that include no really isolated nesting sites, have evolved the ability to relocate their colonies and lay new eggs if disturbed; terns that nest on dunes or sandspits are very sensitive to disturbance and move elsewhere if bothered often, but to compensate for the somewhat vulnerable location of their colonies, they are very aggressive, and will swoop down and strike any large intruder until it leaves. Shearwaters and albatrosses, which cannot lay another egg if the first one is lost, will not abandon a colony site, but are therefore forced to nest only on remote cliffs or islands where there are no predators. While nesting in a group makes eggs and young more vulnerable, the chance of any particular egg or chick being taken is reduced because a predator has so many to choose from. To further reduce the effect of predation, colonies are often highly synchronized, with all the eggs being laid and chicks hatching within a few days of one another; this shortens the total length of time the colony is vulnerable.

Some birds form loose colonies that combine ecological aspects of both colonial and territorial nesting birds. These usually occur in areas where food is exceptionally rich: small islands in the Florida Keys, for example, may contain as many as 14 Osprey nests; the Ospreys defend only the area immediately around their own nests, feeding in the waters of Florida Bay. Unlike truly colonial birds, the Ospreys feed separately and do not react to disturbance as a group. Where fish are less abundant, Ospreys nest farther from one another, with no change in their basic behavior; truly colonial nesting birds cannot breed apart from each other — they seem to require the stimulation of a group, and small colonies are less successful than large ones. Dickcissels, Bobolinks, and other grassland birds sometimes breed on small, clustered territories in one field rather than dispersing evenly over all the available habitat. Although each bird feeds and nests within its own territory, their clustering gives the appearance of a colony.

Pair Bond and Courtship

In all birds it is essential that male and female come together at some time for copulation, so that the eggs the female lays will be fertile, but, as we shall see, whether the pair will remain together to raise the young, the number of mates each bird seeks, and where mating takes place vary with the ecological requirements of each species. In all species studied, however, it is the female that selects the mate, usually on the basis of the male's displays or the attractiveness of the territory he claims.

For most birds, the bond between male and female is established on the male's territory, whether nesting later takes place there or not. Some birds, like the ducks that nest in the high Arctic, will have formed pairs the winter before. Antarctic penguins, faced with the same problem of a short favorable season, mate before returning to the nesting grounds. Similarly, some desert and rainforest birds establish their pair bond before environmental conditions are right for nesting.

A female newly arriving on the territory of any unmated male may at first be treated as another male, especially if their plumage is similar, until she indicates by her behavior that she is not; the male then adjusts his behavior from a display of aggression to one of courtship. The usual features of courtship are vocalizations and displays. In some species like the American Robin, the signs of courtship, other than singing, are practically invisible; in other birds they are very distinctive: male hummingbirds perform special swoops and dives in front of the female; male ducks bob and stretch, each species with a distinctive courting behavior. As mentioned before, the bright plumages and specific courting behavior of male ducks have evolved because most female ducks, plumaged in similar camouflaging patterns, cannot be told apart by the males; the females use the behavior and plumage of the males to recognize their own species.

Three postures assumed by displaying pairs of Western Grebes. The pair on the right is "racing," which involves running over the water and finally diving; the male (rightmost bird) holds his neck in a more contorted position than does the female.

Some birds have displays in which both sexes participate. As shown opposite, pairs of Western Grebes face one another in the water, throw their heads back, and then rise out of the water, running together along the surface side by side in an almost vertical position. Finally the pair, still erect, circle one another breast to breast. Whooping and Sandhill Cranes bow before each other, jump into the air with their wings outstretched, turn, and bow again. Peregrines and other falcons chase one another and stage mock attacks; often food is passed from the talons of the male to those of the female while both are flying at full speed at the end of a dive — all this is done with much screaming.

Feeding of the female by the male is a part of courtship among many species in which both parents help raise the young. Females being fed usually adopt a begging posture and give call notes similar to that of fledglings; some even assume the begging posture when the male has no food to give. Feeding may continue after courtship has resulted in copulation — some males bring food to the female while she covers the eggs or young on the nest. While the gift of food by the male has symbolic value in strengthening the pair bond, it also has real ecological value in supplementing the female's diet when she needs extra food to form eggs, and later when she is unable to leave the nest to feed herself.

The bowerbirds of Australia are so named because the male builds a bower-like structure of leaves and twigs, then decorates it with bright objects such as flower petals, pebbles, fruit, shells, or stolen manmade shiny objects of certain colors. The male displays throughout the breeding season, but after copulation takes no part in nesting.

Group courtship displays are made by males of many species where the male does not participate in nesting. They gather for these displays in places often called "arenas" or "leks," and the same location may be used for many generations. Sage Grouse and prairie chickens strut and dance on open patches in the prairie, making noises by inflating and deflating air sacs on their necks. Females come to inspect the group, select a particular male, copulate, and depart. In New Guinea elaborately feathered male birds of paradise such as the Greater Bird of Paradise shown on page 43 similarly gather in the forest to display their plumes, often hanging upside down. In South America, males of certain hummingbirds gather and perform dances in the air.

Dances of the Whooping Crane. Pairs of Whooping Cranes begin "dancing" in late December, while still on their wintering ground. With young born the previous year looking on, adults bow and leap before one another, wings flapping. The male sometimes leaps over the female's back, and may continue to leap after she has stopped and the young bird has lost interest.

The leks of the Ruff, an Old World sandpiper, are remarkable for several reasons. The Ruff is the only bird in which every male has a slightly different plumage; each has a large "ruff" of feathers around the neck, either mainly dark brown or light gray and white. At leks, dark males are highly aggressive and each defends a display area, while light males are not aggressive and do not have fixed territories. When females visit the lek, some are attracted to males of each behavioral type. The most popular are males that seem to be neither too aggressive nor totally lacking in display territory. In all lek species any female may mate with several males at one lek or may visit other leks.

Birds have evolved several types of mating systems reflecting the most efficient uses of time and energy by male and female parents in differing environments. For the vast majority of birds, the way to produce the greatest number of surviving young is for both parents to work together to raise them. Most birds therefore have only one mate at a time and share parental responsibilities, but how they are divided varies from species to species. In some the female builds the nest, in some the male does, and in some species both work together. Incubation may be evenly shared, or primarily or exclusively the duty of one sex. Among passerines with more brightly plumaged males, the female often does all the incubation — since the male might make the nest visible to predators — while he defends

the territory, watches for predators, and, in some species, brings food to the female. When the young have hatched, it usually requires the work of both parents to keep them adequately fed. Only in environments where food is especially abundant or easily located are young normally raised by one parent.

Most monogamous birds remain together for only one breeding season, although some may nest together in following years because both have returned to the same territory, but the loyalty is to the territory, not the mate — if a new possible mate is found before the old partner returns, the new prospect will be accepted. Some nonmigrants, like Wrentits, titmice, Mockingbirds, and jays often have the same mate more than one year, but these may also be cases of site loyalty — the life span of most small birds is not long enough to give permanent mating systems any particular advantage.

Among longer lived birds such as geese, swans, raptors, and many seabirds, where there is an advantage to faithfulness, pairs are often maintained as long as both partners live, just as their nests are often in the same location or in the same territory year after year — studies have shown that these birds are more successful in raising young when mated to a bird with which they have had at least one year's previous experience. They are all species whose nesting success depends on highly developed feeding skills or, for the geese and swans, whose young feed themselves, extreme vigilance to protect them from a wide variety of predators. Long-term familiarity with a mate who of course shares a long-term familiarity with the territory or feeding localities may improve the chances of a successful nesting.

In a few monogamous birds that raise more than one brood per nesting season the female may switch mates between broods; this is advantageous if it shortens the time she must spend with the first brood, so that she can begin raising a second. Female Eastern Bluebirds and House Wrens sometimes begin nesting with a second male while the first finishes raising the earlier brood. For these birds, two parents are evidently necessary only in the earlier stages; once the young reach a certain age, one is enough.

The briefest kind of monogamous pairing system is found in birds that come together only for mating but leave the raising of young entirely to the female. This is the case with most ducks, where the male spends a long time courting one female but loses all interest when she begins laying eggs. Certain hummingbirds and sandpipers may have similar systems, although males may sometimes later mate with other females.

The regular formation of a pair bond with more than one female, as opposed to indiscriminate mating at a lek, is called polygyny, but it is rare in birds because few environmental conditions exist that make it a reproductive strategy superior to monogamy. A male with more than one mate might bring some food to each nest or might spend all his time defending his territory and females against other

Male Sage Grouse display by spreading their tail, expanding air sacs on the neck, drawing head toward the back, and dropping the wings. In this posture they strut in circles, stoop, make sudden rushes, stamp on the ground, leap over each other, and let air out of the sacs to create a hollow booming sound. After the display reaches a frenzied peak, the males suddenly calm down; females watch the entire display calmly.

males, which is of little direct benefit to the female raising young. Polygyny will only arise where the female can raise her young essentially unaided and where there is some positive advantage to her joining a male who may have other mates.

Nearly all fourteen regularly polygynous North American birds breed in marshes or meadows, where both plant and animal food are abundant enough for a female to raise young singlehandedly. Polygyny occurs in this habitat because a small number of males can keep a large percentage of the best feeding or nesting habitat, so that a female has a better chance of raising young if she is one of several mates of a male holding a good territory than if she is the only mate of, or even helped by, a male with a poorer territory. In the Red-winged Blackbird, which sometimes has up to four mates, there is always a surplus of males occupying the poorer habitat with one or perhaps no mate. The male's territory size does not depend on the number of mates, but on whether they feed within his defended area or just use it to place their nests. A desirable male's territory may be small, but extremely safe from predators since, like most colonial birds, redwings may do most of their feeding away from the nest area. Some of the other regularly polygynous North American birds are the Marsh Hawk, Long-billed Marsh Wren, meadowlarks, Yellow-headed Blackbird, and Swamp Sparrow. The Winter Wren, a woodland breeder, is polygynous only when food is exceptionally abundant.

Polyandry, the mating of one female with two or more males, is a response to different environmental circumstances, occurring in birds whose young feed themselves from the time they hatch, so the major energy strain on the female is not in finding food for the young, but earlier, in finding food for herself to lay eggs which can produce chicks strong enough to run and feed themselves at hatching; parental responsibilities after hatching are only to guard the chicks until they are old enough to fly and detect danger. Given this set of circumstances, polyandry may occur in species where a female can leave more offspring if she mates first with one male, lays a set of eggs for him to raise, and then does so with a second and possibly a third male, rather than laying only one clutch and raising the young herself. The success of this strategy depends on abundant food for the female when she is forming her eggs. Producing eggs is a considerable energetic strain, and few birds can produce more than one set quickly, so polyandry is common only in a few tropical families where a long breeding season lessens the strain on the female. In North America, where all the eggs must be laid during a relatively short time, polyandry is definitely known only for the Spotted Sandpiper: female Spotted Sandpipers may lay eggs for up to four males, sharing the incubation of the last-laid set.

Mating systems which may lead to polyandry can be seen in other shorebirds: female Sanderlings lay two clutches — one is incubated by the male, the other by the female; "sexual reversal," where larger, more brightly colored females pursue males who build a nest and take over all responsibilities after eggs have been laid, is well known in the phalaropes, although it has not been shown that females of any phalarope species regularly lay eggs for more than one male.

Promiscuity, where each male may copulate with any number of females and each female with more than one male, is the standard mating system for species where the two sexes encounter one another only when the male is displaying to attract a mate. Lek species are definitely promiscuous, but many species that display singly like the Ruffed Grouse and American Woodcock may be as well.

The Nest

By the time a female is ready to lay her eggs, a nest must be ready to contain them. The purpose of the nest is to hold and shelter the eggs, and for most birds the young, until they are mature enough to fly away. Most small birds try to conceal their nests, making them harder for predators to find and perhaps protecting them from bad weather as well. These nests are easiest to locate while they are being built; look for birds carrying twigs, bits of grass, or other material in their bill, but don't follow so closely that you scare them away from their chosen nest site or upset their work.

Every species builds a distinctive kind of nest. Some are fairly flexible about location or materials used, while others will not inhabit an area that lacks specific requirements. The Parula Warbler will not breed where it cannot hang its nest in *Usnea,* a northern lichen that grows on trees, or *Tillandsia,* the Spanish Moss of the South. With practice, you will be able to recognize most of the bird nests in your area, basing your judgment on materials, location, and size.

The location of the nest is usually chosen by the female. Males often lead females to what they consider suitable sites, but the female usually makes the final decision. In the Scops Owl, an Old World species closely related to the Screech Owl, the male shows the female possible nesting cavities by perching in the entrance, spreading his wings, sticking his head into the hole, and drumming on the wall with his beak. In some wrens the male chooses a hole and begins filling it with nest material, and the female then takes over and finishes the job. Among Cedar Waxwings, the nest site seems to be chosen jointly.

There are several ways, both active and passive, that birds can protect their nests from predators. The nest may be in a place difficult to reach, as is an oriole's nest at the tip of a thin twig, or hidden in a hole in a tree, rock, or bank. It may be part of a large construction in which the actual nest chamber is hard to find: Black-billed Magpies create a huge mass of sticks with two entrances and a nest in the center. The nest can be camouflaged by placing material over the eggs when they are not covered by a parent, as is done by the Pied-billed Grebe, or by resembling other objects in the habitat: the Ruby-throated Hummingbird's lichen-covered nest resembles a branch stub. The eggs and young birds themselves sometimes have camouflaging patterns. The nest may be defended by the adults, or be defended incidentally by other animals nearby; Brant in northeast Greenland nest near Gyrfalcons, and Common Grackles sometimes nest on the edge of an Osprey's nest; both hawks feed on other items and scare predators away from the area of their own nests. Finally, the parents may try to mislead a predator: a Killdeer pretends to have a broken wing, dragging it on the ground, giving distress calls, and leading a predator far from the nest before it flies off.

A few birds build no nest at all. The Whip-poor-will lays its eggs on the forest floor. The White Tern (page 130) is famous for placing its single egg on a horizontal branch, from which it never seems to roll. Emperor Penguins, which live on the Antarctic ice, keep their single egg on top of their feet, covered by a fold of belly skin. Murres put their egg on a flat cliff ledge.

The most minimal nests are made by Least Terns and certain shorebirds that may scrape out a little hollow in the sand and surround their eggs with a few pebbles, shells, or bits of grass. Anything more would make these eggs, which are colored to blend in with their background, easier to find.

The majority of birds build nests of sticks, grasses, or plant fibers. The familiar, open-topped cup shape is commonest. Various species place the nest on the ground, in a shrub just above it, or anywhere up to the top of the highest tree available. Nests off the ground are usually safer from predators or flooding, but elevated sites are of course not available in every habitat. Surprisingly, many warblers and other birds that feed high in trees always place their very well concealed nests on the ground, so a ground location must still have certain advantages. Song Sparrows place their first nest of the season on the ground and later ones in bushes, which by then are covered with leaves, and, similarly, American Robins often nest first in evergreens and later in deciduous trees, which did not have leaves when the first nest was being built.

Cup-shaped nests in bushes or trees are usually placed in a branch crotch, against the trunk, or on a flat branch. The nest may be very simple: Mourning Doves build a platform of twigs so thin that the eggs may be seen from below. Cuckoos, Gray Catbirds, and Cardinals build more substantial twig nests. Most passerines make cups with layers of increasingly fine material, using twigs to create the shape, filling in with grasses, and lining the inside with soft plant fibers; many sparrows use animal hair in the lining and other birds specialize in stray feathers.

Larger birds build mainly with sticks. Egrets and herons make a simple platform, but sometimes use the nest again and add new material in later years. Hawks and owls often use the old stick nest of a Common Crow, or build a similar one of their own. Ospreys and Bald Eagles create huge piles of sticks, often at the top of exposed dead trees, adding to them each year until they collapse of their own weight; Bald Eagles often have two large nests within their territory, using each in alternate years to reduce the number of nest parasites.

Domed or hanging nests are rarer than simple cup-shaped ones. Meadowlarks and a few other ground nesters build a roof that covers the eggs or young while the parent is off the nest, adding to their protection. (This is the source of the Ovenbird's name.) Long-billed Marsh Wrens build covered nests attached to reeds in marshes. Hanging, woven nests are made by vireos and orioles; the vireo nests are cups suspended from their rims, but the Northern Oriole weaves a hanging pouch, with the eggs laid at the expanded bottom of the six-inch sac. Oriole nests are usually placed high in a tree at the tip of a thin twig, where they are easy to see but safe from all predators.

Mud is such an essential nest ingredient for some birds that if

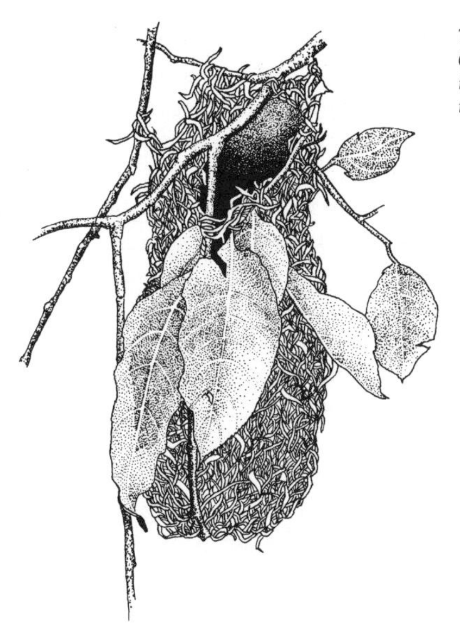

The pouch-shaped nest of a Northern Oriole is always hung from the thinnest twigs, often on a branch over water or a roadway.

none is available robins and Barn Swallows may make it by dipping their feet in water and then standing in dry earth. The American Robin uses mud for the middle layer of its nest; while building, it sits inside the nest, rotating its belly to create a smooth round cup of the right size. The mud layer is then lined with fine plant material. Grackles and several thrushes build similar nests.

Swallows use mud to create nests of different shapes; two are illustrated above. The Barn Swallow builds a cup which may adhere by itself to a vertical surface, while Cliff Swallows make covered nests with a spout-like entrance on the side — both now use manmade structures much more often than caves or cliffs. The Cliff Swallow usually nests under an eave; it is declining in some areas because more people now paint their barns and the birds cannot make mud stick to the paint.

Hornbills, large African and south Asian birds, nest in a tree hole and use mud to plaster up the entrance. The female remains in the hole throughout the incubation period and while the chicks are young; food is passed to her by the male through a narrow hole left after the plastering.

Holes in trees are the nest site for many other birds. Woodpeckers are the best-known excavators of holes, but nuthatches, chickadees, and titmice can dig their own holes if the wood is soft or rotten. The many hole-nesters that cannot dig their own holes depend on woodpeckers and natural tree cavities of the right size. In the deserts of the Southwest, woodpeckers dig holes in large cacti and other birds use them later. The ecological advantage of holes is that they are an extremely safe nest site, with protection from the weather and from most predators. This safety lessens the pressure on birds to raise their young to the stage of flight and independence as quickly as possible; because young are safest in the hole, they leave their nest at a comparatively later stage of growth than the young in open nests. Rather than feeding their young the maximum amount so they will grow as quickly as possible, hole-nesters can use the same amount of food to feed more young at a slower rate. In fact, the largest passerine clutches are laid by the hole-nesting Blue Tit, a European chickadee; clutches average 12 or 13, with highs of 18 and 19.

However, good holes are scarce. The number of species that depends on holes is small and often their population is kept lower than the food supply permits by a shortage of holes. Birds that depend on woodpecker holes suffer seriously if anything lowers woodpecker

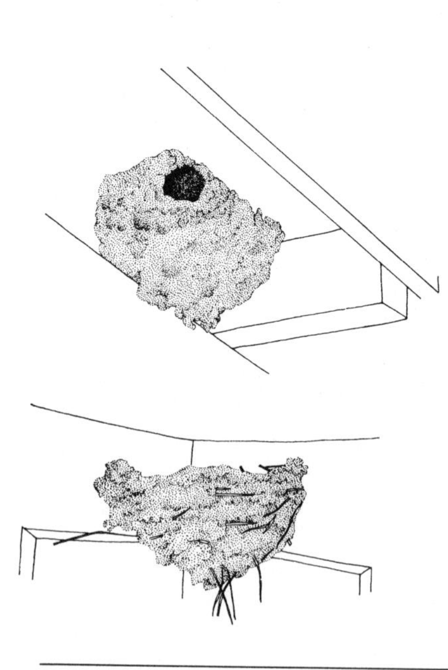

Differing styles in swallow architecture: the enclosed nest of the Cliff Swallow, usually attached to an overhanging ledge, and the cup-shaped nest of the Barn Swallow.

Nest of the Long-billed Marsh Wren.

populations. Birdhouses can increase the density and the populations of many hole-nesters.

Holes or tunnels in the earth are used by a few species, but soil conditions limit these sites and the numbers of species that can use them. Belted Kingfishers and Bank Swallows dig long tunnels into sand or gravel banks; at the end of the tunnel, 4 to 5 feet long for the kingfisher, 2 to 3 feet long for the Bank Swallow, a slight enlargement forms the nest. Petrels, shearwaters, and puffins dig shorter tunnels into hillsides for their nests. Burrowing Owls often use old prairie dog or ground squirrel tunnels. In the tropics, some birds dig into termite hills, deriving some protection from the insects.

Marsh nesters like the Red-winged Blackbird and Long-billed Marsh Wren weave their nests into the reeds. Larger birds, like rails, make a platform of marsh vegetation in a reed bed over the water. The tops of muskrat and beaver lodges are used by some birds, but grebes build floating nests of soggy vegetation anchored to growing plants.

Some birds require special materials for their nests: Yellow Warblers and American Goldfinches use plant down from willows, poplars, and thistles to make their nests soft and waterproof; Ruby-throated Hummingbirds use spider webs to hold the nest together, covering the outside with camouflaging bits of lichen; flycatchers of the genus *Myiarchus,* which includes the Great Crested Flycatcher, always include an old snakeskin in their nests (in recent years some individuals have substituted a piece of plastic for the snakeskin).

Swifts glue a platform of twigs together with a secretion from the salivary glands. In the East Indies there are swifts whose nest is made almost entirely of these secretions; natives use the nest to make a soup, which can now be bought in certain oriental food markets in America. Many birds use their own or found feathers to line the nest. Ducks and geese grow a special down that they pluck for the nest; the nest down of eiders is the most famous as an insulating material in coats and sleeping bags.

Feet, body, and bill are used in nest construction. Small birds usually carry material to the nest in their bill, but larger birds with sticks may use their feet. Scraping a hollow in the sand, the first step for nests of gulls, terns, and some shorebirds, is done with the feet; the bill is used for precise placement of twigs, grasses, and for all weaving. Moving the body around in the nest brings it to the right shape.

Time spent constructing the nest depends on how elaborate it is. American Robins take from 6 to 20 days, most individuals taking closer to 6, while Song Sparrows usually need only 4 or 5 days. In the tropics, where the breeding season is long, construction may be two or three times slower than it is for a comparable nest in the temperate region, but in the Arctic, passerines sometimes save time by reusing old nests. Birds usually work only a few hours a day on the nest. A few, like doves and herons, continue adding sticks after they have laid eggs and begun incubation.

The role of the sexes in nest construction varies from group to group. Among hole-diggers like woodpeckers, kingfishers, and Bank Swallows, males and females share the work. Males of several wrens such as the Long-billed Marsh Wren build dummy nests while the female finishes the real one; these dummies may be used both to confuse predators and as roosts for the male. Among pigeons and doves, the female sits on the nest and the male brings her all the material; it is the reverse with frigatebirds, where the male builds with what the female gathers. In Ravens, both sexes gather, but only the female builds. Females that raise their young unaided by the male naturally build the nest alone, as do Red-eyed Vireos, Ovenbirds, and probably other warbler species in which both sexes raise the young. The nest is built entirely by the male in Phainopeplas and certain weaver finches, and by species like the Ostrich and phalaropes where the male raises the young unaided.

Group nests are built by a very few birds. The Sociable Weaver, an African weaver finch, nests in colonies from about ten to hundreds of pairs, building a huge structure of straw in a large tree; each pair of weavers has its own entrance and nest cavity in the structure. The communal structure helps maintain an even temperature in the nests throughout the year, which is important to the weavers since their nesting schedule is associated with rainfall, not season, and may occur in winter when outside temperatures are occasionally below freezing.

Smooth-billed Anis, a tropical American cuckoo, live in groups of up to twenty birds that defend a territory and together build a large stick platform in which the group's females each lay up to seven eggs. Layers of about eight eggs are separated by leaves; one bird of either sex incubates at a time, and when the young hatch they are fed by all members of the group. If a nest is robbed, it is covered with leaves and a new one is started immediately.

The copulatory stance of Common Terns.

The most remarkable nests are made by the megapodes, a family of ten chicken-like species inhabiting Australia and the islands off southeast Asia. Male megapodes spend the entire year tending a nest that is a pile of sand, warm volcanic ash, or rotting vegetation into which several females lay eggs that the male then buries in the pile. The heat of the sun, the volcanic ash, or the decomposing vegetation keeps the pile warm and incubates the eggs, with the male adjusting the amount of material on the nest to keep the temperature even. When the young megapode hatches, it digs its way out of the pile and goes off to live on its own, while the male continues to tend the nest; young megapodes hatch fully feathered and capable of flight.

Eggs

The eggs inside the oviduct of the females are fertilized by the male during copulation, which may take place after a complex series of courtship displays or may occur with no preliminaries. The female usually invites the male to copulate, by giving a particular call or assuming a certain posture.

Copulation is usually repeated frequently during the first part of the breeding season, before and during the egg-laying period. Some birds such as ducks and House Sparrows may begin copulating in midwinter, long before eggs are laid, but other birds have a very brief period of interest in copulation; females of lek species may visit the males only once or twice before retiring to nest.

When the egg has been fertilized, it moves down the oviduct. The center of the egg, the follicle, is mainly yolk, and is covered by a thick layer of albumen, the "white," from glands in the middle of the oviduct. The amount of yolk varies in different birds: the eggs of birds that are helpless when they hatch are about 20 percent yolk; eggs of birds whose young can run about and feed themselves at hatching, like chickens, are up to 35 percent, and the megapode egg, which must nourish a chick that hatches fully feathered, has a yolk that takes up 70 percent of the egg. At the end of the oviduct the egg receives a shell from glands in the uterine portion of the tract, and the pigments which put various colors or marks on the shell. For most birds, production of an egg takes 24 hours, with one egg cell released from the ovary each day: addition of the egg white surrounding the yolk requires about 3 hours, inner shell membranes are added in about 1¼ hours, and the hard shell and application of pigment take from 18 to 20 hours.

Eggs occur in a wide variety of sizes, shapes, and colors. Even in a box of a dozen hen's eggs you will find some larger or more pointed than others. The color, markings, shape, and size of bird eggs are generally consistent within each species, because these characteristics serve specific ecological functions. Egg shape is partially a reflec-

Eggs of different shapes and with different marks: from left to right, Double-crested Cormorant, Common Murre, Great Horned Owl, Least Tern.

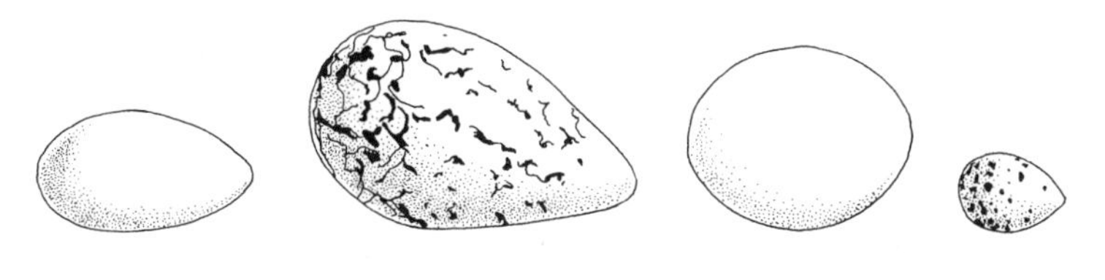

tion of nest site: murre eggs, placed without a nest on narrow cliff ledges, are extremely pointed at one end, so that if pushed, they roll in a circle around the narrow end rather than rolling straight, like a ball, off the cliff. Pointed eggs are also laid by shorebirds with regular clutches of three or four; these can be most easily arranged, point-down, under the incubating bird. Eggs almost perfectly round are laid by hole-nesting birds, since these eggs cannot roll out of the nest.

Since reptile eggs are white, probably the first bird eggs were also white. Although now bird eggs occur in many colors and with many kinds of markings, the basic color of most is still light, with white or off-white commonest; robin's-egg blue is well known because it is a fairly unusual color. A few other thrushes lay unspotted blue eggs, and Gray Catbirds lay unmarked bluish-green eggs. Glossy, porcelain-textured eggs of blue, green, yellow, purple, or brown are laid by various species of tinamou, quail-like birds of Central and South America; the functions of their unusual textures and colors are not known. The eggs of many birds have spots or scrawls on them; the pigmented markings over the basic color of the egg form scrawls or scratches if made while the egg is moving down the oviduct, and make spots if applied while the egg is not moving.

In general, egg colors and markings serve an ecological purpose: nearly all ground nesters have spotted eggs that blend in with the area around the nest; birds that nest in somewhat concealed places, in bushes, or in trees, often have eggs with a pale bluish or greenish background, marked or unmarked — these colors and markings may make the eggs less visible in the dappled, shady light inside a bush; eggs laid in dark holes or burrows are usually white, so they will be easily seen by the parents. There are, however, exceptions to all of these rules. Whip-poor-wills and Chuck-will's-widows lay highly visible, slightly spotted, light eggs on the forest floor, but they are almost always covered by the very well camouflaged parents.

In a few species, eggs change color with age; rheas, ostrich-like birds of the South American grasslands, lay green eggs which during incubation successively become yellow, blue, and finally white at the time of hatching.

Colors and markings are usually consistent within a species, but a few birds like Royal Terns and Common Murres that nest in dense colonies where, unlike other birds, they recognize their own egg, produce eggs with a wide variety of markings.

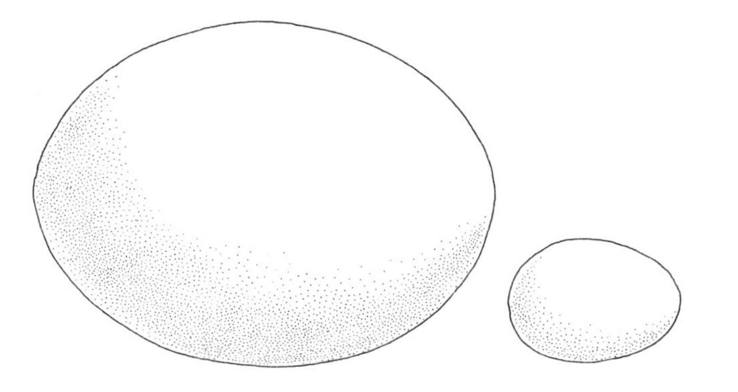

The size range of bird eggs: from left to right, Ostrich, domestic hen, Ruby-throated Hummingbird.

Eggshell textures also vary. The glossy tinamou eggs are unusual, but most small birds have eggs smoother than a hen's. The eggs of ducks and some other water birds are greasy and water resistant, but grebe eggs, which are often placed in nests of soggy vegetation, are chalky. Eggshell thickness is related to the overall size of the egg.

Egg size varies from the half-inch-long eggs of hummingbirds to the six- to eight-inch-long eggs of Ostriches. (The largest bird eggs ever known were laid by the now extinct elephant birds of Madagascar; the birds weighed a thousand pounds and their eggs measured 13 x 9.5 inches and weighed 18 pounds.) However, birds of the same size do not all lay eggs of the same size — egg size reflects the level of maturity of the chick when it hatches. Some birds, including chickens, gamebirds, ducks, sandpipers, and plovers, are born with a coat of down, can run or swim, and are able to feed themselves; they are called "precocial." Most birds are born helpless, sometimes blind and without down, and are fed by their parents until mature enough to fly and feed themselves; these are called "altricial." Because precocial birds must have done more of their growing before they hatch than altricial birds, they require a proportionately larger egg, with more yolk, than altricial birds — the drawing on page 156 shows how although the American Golden Plover and Eastern Meadowlark are each nine inches long as adults, the plover's eggs are 1.9 x 1.3 inches while the meadowlark's are 1.1 x 0.8; the plover chick can run and feed itself a few hours after hatching, while the meadowlark is born completely helpless.

Most birds lay eggs that are consistently the same size. A long-term study of Arctic Terns, however, found that individuals laid larger eggs when 6 to 8 years old than they did when younger or older. For another unknown reason, in certain gulls that regularly lay three eggs, the last egg laid is narrower, but not shorter than the others; normally all three chicks are raised to maturity. The Red-footed Falcon, an Old World species, lays three or four eggs, the first always smaller and paler than the last; all eggs normally hatch. Penguins of the genus *Eudyptes* lay two eggs, the second approximately one-third heavier than the first, and usually only the second, larger, egg is incubated. One of the five *Eudyptes* species incubates both eggs, but only takes care of the chick from the smaller egg if the other doesn't hatch. Either chick ultimately grows to the normal adult size. For the species that incubates both eggs, the smaller one serves as a reserve; in the other penguins the small egg is only a relict from the time when both were incubated.

The altricial Eastern Meadowlark lays a smaller egg than the precocial American Golden Plover, although adults are the same size.

The normal number of eggs laid in one nest is called a clutch. Clutch size varies from one to about twenty in different species; it is influenced by several environmental factors. In general, birds produce many fewer eggs than the thousands normally laid by insects and some fishes, or the hundreds or tens laid by amphibians and reptiles, because birds give more care to their eggs and young, assuring that a higher percentage will survive. The largest clutches are still laid by birds whose eggs and young face the most dangers; species with very safe nest sites and few enemies of young or adults usually lay smaller clutches. As we noted, the number of eggs laid by species that feed their young is the greatest number that can, on the average, be fed. The number of eggs laid by precocial species, on the other hand, is determined by the amount of food available to the female at the time of laying. For some species both factors may be involved: Snowy Owls normally lay 4 or 5 eggs, but will lay as many as 8 or 9 in years when an abundance of lemmings, their principal prey, provides enough food for the female to form that many eggs and later feed all the chicks; Pomarine Jaegers, which also feed on lemmings, do not nest at all in poor lemming years.

Some birds lay a fixed number of eggs and if these are lost cannot lay again until the next breeding season. Among these are all the tubenoses, which never lay more than one egg. The egg they lay is proportionately rather large, it takes a very long time to hatch compared with other eggs of its size, and the chick grows very slowly. To produce such an egg requires a lot of effort by the female, presumably too much for her to make twice in one year; another reason for not laying again is that since incubation and caring for the young take so long, a chick from a second egg laid later would probably not be mature by the end of the nesting season, and the energy spent in raising it would be wasted. Limited food supplies are probably why tubenoses can raise only one young and why its growth rate is so slow. In experiments with various albatrosses, shearwaters, and petrels where two young were placed in the same nest, both starved.

Other birds always lay the same number of eggs in each clutch, but can produce a second or even a third clutch if earlier ones are lost. Every species of pigeon lays either one or two eggs, but the domestic pigeon may nest throughout the year, and Mourning Doves, which always lay two eggs, raise several broods each year. Similarly, every species of hummingbird lays two eggs, although hummingbirds occur from Alaska and high in the Andes to the lush tropics and food resources are different in their variety of habitats, so that some species might be able to raise more than the usual two young. Rather than evolving different clutch sizes, hummingbirds have evolved different growth rates — where food is plentiful, young hummingbirds grow faster than where it is scarce. Some, like the Black-chinned Hummingbird of western North America, raise two or three broods in the nearly three months it takes other species to raise one.

Most birds have a clutch size that varies by two or three. American Robins, for example, most often lay 4 eggs, but some individuals more or less regularly lay 3 or 5. Four or 5 is the usual clutch size for most passerines, although titmice, kinglets, nuthatches, and creepers regularly lay 6 to 10. European titmice sometimes lay as many as 20 eggs. Various raptor species lay from 2 to 5 eggs. Ducks and game birds, whose eggs and young face many dangers, have clutches of 8 to 15.

Most passerines and other small birds lay one egg a day until the clutch is complete. Often the egg is regularly laid at a particular time of day like early morning. Larger birds usually lay an egg every other day; some require even longer to form each egg.

If the first clutch is lost, some birds are able to lay again almost immediately. This is an important adaptation for birds like terns that nest in places where the eggs may be destroyed by flood. It also allows terns to move to another area if disturbed, without losing an entire nesting season. Their incubation period is shorter and the growth rate faster than in tubenoses of the same size, so that late-starting terns may still have enough time to raise their young. The usual clutch size of 2 or 3 in most terns indicates that their food is also easier to find. Ducks and hens, on which man depends for eggs, can also lay many more eggs than the normal clutch size, but some wild birds have the same potential: an experiment in removing eggs from a Common Flicker's nest led the female to lay a new egg each day for 71 days.

Birds that lay a variable number of eggs are subject to other environmental influences, such as day length and food abundance, on clutch size. Species that breed over a wide north-south range generally produce larger clutches in the northern portion of their range, because with more hours of daylight they have time to feed more young. Within one family, temperate or arctic nesters usually lay more eggs than do their tropical relatives. However, since there is probably no shortage of food in tropical environments, the reason so many tropical species have small clutch sizes is not clear, but the principle of birds laying as many eggs as can be raised, a theory that grew from work with temperate zone birds, does not seem to apply in the tropics. The fact that island populations of birds with variable clutch size lay fewer eggs than do mainland populations of the same species may mean there is less food available on the islands.

The number of broods raised may also vary within a species. Most larger birds grow too slowly for more than one brood to be raised in a single breeding season, but many passerines and other small birds can raise two broods: Barn Swallows, House Wrens, Mockingbirds, American Robins, bluebirds, and Cardinals are among the birds that regularly attempt two broods. In contrast with the number of eggs laid per clutch, which when variation exists is higher in the north, a larger number of broods is raised in the southern part of a bird's range, where the breeding season is longer — Mourning Doves in the southern United States may raise four broods between February and October, while northern populations raise only two or three in a shorter season.

Few birds have the ability to recognize their own eggs. Under normal circumstances, birds need only recognize their nest or the nest site. For most colonial birds it is simpler to return to the nest by remembering where it is located than by recognizing the eggs in it. Only birds that nest in very dense colonies where there are no landmarks to distinguish one nest location from another can recognize their own egg. Royal Terns and Common Murres nest in colonies of this type and have evolved very individually marked eggs that make recognition simpler; each lays only one egg, further simplifying recognition. In experiments, Royal Terns were able to recognize their own eggs when moved to the adjacent nest; recognition was mainly by the color and markings of the egg, but site and recognition of neighboring adults also helped.

Most birds will accept any solid object in the nest as an egg. Experiments with gulls and terns have shown that they will incubate objects the size, shape, and pattern of their eggs, cubes patterned like their eggs, and eggs or round objects in different sizes and colors; only red objects cause some concern and are usually removed from the nest. Other experiments have shown that birds are especially stimulated by egg-like objects larger than their own. Oystercatchers will leave their own eggs to incubate an imitation two or three times larger, and redpolls will attempt to incubate a hen's egg or a tennis ball. Occasionally, in natural, non-experimental situations birds are stimulated to incubate objects resembling their eggs — gulls and terns may incubate an appropriately sized and marked stone that has rolled into the nest. Emus, ostrich-like birds of Australia, have been found incubating paddy-melons, which are the same size and color as their eggs but a different shape; because the melons don't fall off the vine until ripe, when they no longer resemble Emu eggs, the birds must have broken them off and moved them into the nest.

Incubation

Except for the megapodes, whose eggs are warmed by the sun or rotting vegetation, all birds use their bodies to keep eggs at the high temperature needed for development. A few penguins hold the egg on their feet to prevent it from touching the ice, but all other birds sit on their eggs to warm them. Most bird eggs require a temperature in the nineties Fahrenheit. Embryo development stops when the egg is less than 82°, but eggs can survive varying lengths of time at lower or higher temperatures.

During the nesting season most incubating birds develop an "incubation patch" or "brood patch" which enables them to warm the eggs more effectively. The brood patch is a featherless area on the belly with many blood vessels near the surface of the skin. Down feathers covering this area are lost by molt, rather than plucking. The bird parts the feathers on the sides of the brood patch and applies the skin directly to the eggs, while the feathers surround the eggs. Some birds, like gulls, develop separate brood patches for each egg. Gulls normally lay three eggs; they have three brood patches and cannot incubate four eggs. Brood patches normally develop only in the sex that incubates the eggs. In nearly all passerines only the female has a brood patch, although males of many species incubate.

Incubation usually begins when the clutch is complete, although before then birds often sit or stand over the eggs to protect them, but do not apply the heat needed to begin their development. If incubation begins before the clutch is complete, the eggs will hatch on different days, forcing the parents to leave the nest to find food for chicks at the same time the other eggs must be incubated. Only species in which both parents tend the nest can adopt this strategy, which has advantages for certain birds. Raptors, gulls, terns, and a few other birds whose food varies in abundance begin incubating after the first egg is laid. If there is enough food available when the chicks hatch, all will be fed adequately and all will survive, but if there is not enough food to feed all, the older, larger chicks will get most of the food and will survive; were the chicks all the same age and size, and the food divided equally, there might not be enough for any.

Even after incubation begins, the eggs are left uncovered some of the time while the incubating parent feeds. The amount of uncovered time is less in species where both sexes take turns incubating or where the male brings food to the female on the nest so that she doesn't have to leave. In general, birds may stay away from the nest longer at midday and only briefly during cooler hours; they rarely leave during rain or storms, or at night. Passerines may leave the nest for a few minutes every hour or half hour, while larger birds whose food is harder to find stay away longer, but also have lengthier incubation sessions. Shearwaters and penguins that travel far for their food always have the eggs covered by one of the parents; even though one may be away for as long as two weeks, the other never leaves the nest until its mate returns.

In most birds incubation is done by the female. The males of many species, of course, take no part at all in incubation, but even where they do, the female usually covers the nest for more hours. In the Mourning Dove and some other pigeons, the female incubates at night and is relieved by the male for an eight-hour stretch during the day. Among passerines and other small birds in which both sexes incubate, the daytime changeovers are much more frequent, with the female incubating at night. For the seabirds that are gone for more than 24 hours, incubation shifts are about even. Among species with "sex reversal" such as the Ostrich, Emu, kiwis, tinamous, and phalaropes, where the male is responsible for eggs and young, the female does not incubate at all.

The changeover at the nest is usually quiet and fast, to avoid attracting attention; if the sitting bird is not willing to leave, it may be pushed over by its mate eager to incubate. Returning Common Terns sometimes bring a fish for the mate leaving the nest; still more elaborate changeover ceremonies, involving calls, bobbing, and bowing, occur in shorebirds that nest in the open.

The length of the incubation period depends less on the size of the egg than on the maturity of the chick at hatching: eggs of precocial birds take significantly longer to hatch than do similarly sized eggs of altricial birds. The young of precocial species do more of their maturing within the egg, so that a Mallard chick that hatches after 27 days may be at the same level of maturity as an altricial bird that spent 15 days in the egg and 12 being fed in the nest. Similarly, among altricial birds, those with slower growth rates as chicks also have slower growth rates within the egg, and consequently longer incubation periods. The Least Tern lays an egg weighing 9 grams; it takes 21 days to hatch and 30 days for the chick to reach the stage of strong flight. The Leach's Storm Petrel, in comparison, lays a 10-gram egg that takes 42 days to hatch and 66 days for the chick to mature. Birds with inaccessible nests, where the eggs are safe, often have proportionately longer incubation periods than relatives nesting in less safe sites. Similarly, species nesting in predator-free environments may have longer incubation periods than those whose young are safer as chicks than as eggs. The Emu, which has no enemies, takes 56 days to hatch its 600-gram egg, while the Ostrich, whose eggs are eaten by many animals, hatches its much larger, 1500-gram egg in only 42 days.

The shortest incubation period for birds is ten days. House Sparrow and Red-winged Blackbird eggs sometimes hatch in ten days, but eleven or twelve days is more usual for them and for most small passerines. American Robin eggs usually hatch in thirteen days, the Blue Jay's in seventeen. Hummingbirds, with much smaller eggs but slower growth rates, take proportionately longer. The three-inch Allen's Hummingbird has an incubation period of 15 days, and the 4½-inch Andean Hillstar, a temperate zone South American species, requires 20 or 21 days. The longest incubation periods are found in large seabirds with slow growth rates; the Royal Albatross egg takes 80 days to hatch. The California Condor, in comparison, takes 56 days, and the Ostrich, as noted above, with the largest bird egg, 42.

The environment also affects the incubation period. Broods started later in the season may take slightly less time to hatch, since the air temperature is higher and the incubating bird needs less time off the nest to find food. In the tropics, however, the incubation period of many birds is significantly longer than that of their temperate zone relatives. Some tropical flycatchers take more than 20 days, while North American flycatchers take 13 to 15.

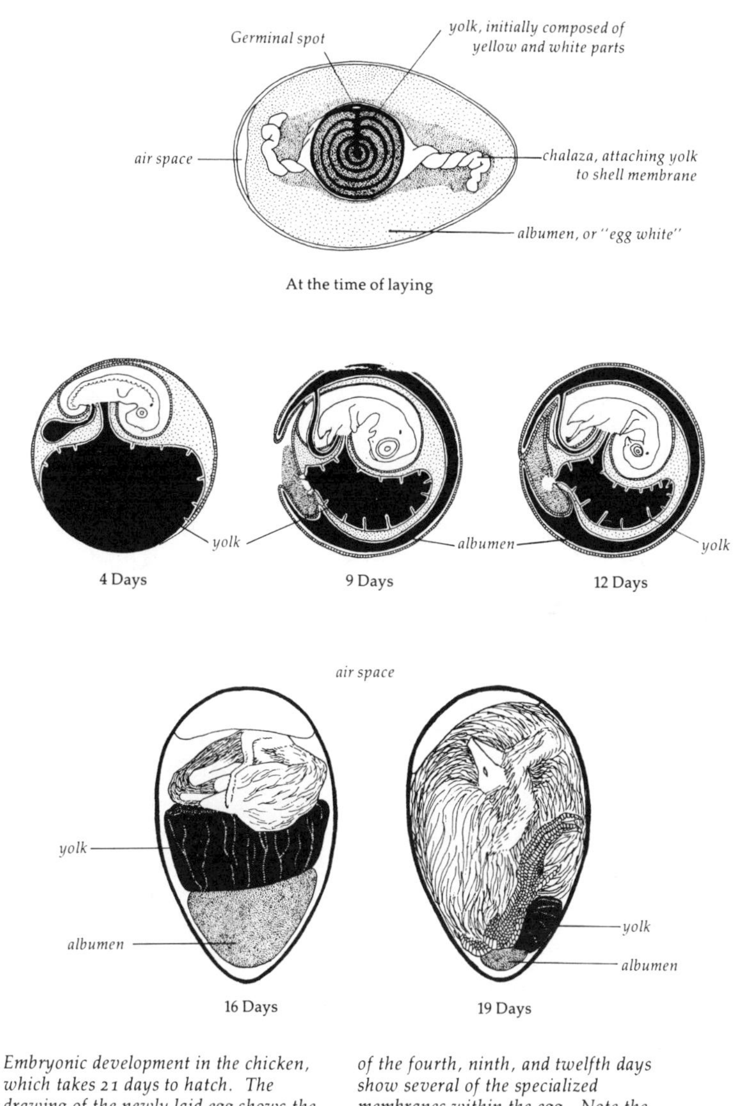

Embryonic development in the chicken, which takes 21 days to hatch. The drawing of the newly laid egg shows the yolk's alternate layers of "yellow yolk" and "white yolk"; the "white yolk" is actually only 3 or 4 percent of the total yolk materials and contains less fat and pigment than the other layers. Drawings of the fourth, ninth, and twelfth days show several of the specialized membranes within the egg. Note the gradual shrinkage of the yolk and of the albumen, or "egg white," which in the ninth day does not entirely surround the embryo (it is not shown in the drawing of the four-day-old embryo).

During incubation most birds regularly turn the eggs with their bill or body to heat them more evenly. Some birds cool or humidify their eggs on hot days by sprinkling them with water brought on the bill, feet, or breast feathers. Eggs may also be shaded on very hot days. The initial rate of development within the egg is rapid: in the chicken, with an incubation period of 21 days, the first signs of the nervous system, digestive tract, and blood vessels appear within 24 hours after incubation has begun; within 36 hours the heart begins beating and food is brought by veins from the yolk to the embryo; in 48 hours eyes, ears, and brain can be seen. Respiration takes place through the shell, which, no matter what the outside texture, is porous enough for the intake of oxygen and the elimination of gaseous wastes. By the end of the incubation period of any bird, the chick almost completely fills the shell. It has used all the egg white and all the yolk except that part attached to its abdomen, called the "yolk sac," which helps nourish it immediately after hatching. Before hatching, the chick thrusts its bill into the airspace at the blunt end of the egg, begins regular breathing, and gradually begins to break out of the shell.

Incubating birds have several different responses to predators. Most small birds can do nothing to stop a predator from robbing their nest, but they may flutter or perch nearby giving calls of alarm. Robins have evolved eggs which are supposedly bad tasting and avoided by some predators. Birds the size of American Robins and Common Grackles will chase a squirrel and perhaps even a cat. Blue Jays are still more aggressive; some will even strike at humans. A few small birds are well known for their aggressiveness to larger birds. Hummingbirds and kingbirds will chase crows and hawks aways from the nest and out of their territory. Large hawks and owls can be very dangerous to humans approaching their nest; there is much individual variation, but some will strike an intruder with their sharp talons, so it is best not to approach their nests closely. Among seabirds, terns are the most aggressive; Common Terns will often dive at and strike an intruder in the colony. Another tern response to repeated disturbance is to simply desert a site and nest elsewhere. Most tubenoses nest in places where they have no mammalian predators and where theft of eggs by gulls or skuas is best avoided by sitting tight on the nest; they show no signs of fear and can be picked off the nest by hand, but some, including the Northern

Fulmar, spit an unpleasant smelling, long lasting oil at anything which comes too close. As mentioned earlier, many shorebirds will pretend to have a broken wing, dragging it and calling plaintively as they lead a predator away from the nest. The Chuck-will's-widow, which builds no nest, will carry its eggs in its bill to a new location if disturbed.

Chapter Nine

A Golden Eagle chick shortly after hatching.

The Breeding Cycle: Hatching and Development of Young

Hatching

At the end of the incubation period, eggs with live embryos hatch; eggs that fail to hatch are either infertile or contain embryos that died at some stage of development. During its last days or hours within the egg, the chick uses its bill to make a series of cracks around the circumference of the shell, usually at the larger end. After "pipping" the shell, it punctures a small hole and then kicks and struggles within the weakened shell, eventually splitting it open. A parent occasionally helps by picking at the opening.

The time spent pipping and hatching varies individually and from species to species: Common Tern chicks may begin pipping the egg two days before they hatch, but grebes, which have floating nests, hatch quickly, three hours from pipping, to lessen the danger of the chicks' drowning should water come into the nest while they are in open shells. Ostriches have a distinctive way of hatching: the chick breaks a hole in the shell, inserts its beak and one foot, and shatters the shell by stretching its body and kicking.

To help break out of the shell, all birds have an "egg tooth," a horny protuberance on the tip of the bill, the rest of which is still relatively soft. The egg tooth is found also in reptiles and some frogs; on birds its placement varies with the shape of the bill. On most birds, it occurs only on the tip of the upper mandible, or if the bill curves downward at the tip as in hawks, just before the curve. Birds with mandibles of equal length, like sandpipers, have an egg tooth on both mandibles. Some birds drop the egg tooth after hatching; in others it is absorbed by the bill. The rate as well as the manner of disappearance varies widely, but the egg tooth can always be seen on a newborn chick, like the Roseate Spoonbill on page 168.

The egg tooth of a newly hatched Roseate Spoonbill is on the tip of its upper mandible.

Whether or not all the chicks hatch at approximately the same time depends, as we have seen, on when incubation begins. Synchronous hatching is especially important for precocial species like grouse and pheasant that leave the nest a few hours or a day after hatching, because the hen may abandon fertile eggs that do not hatch within a few hours of the others. Chicks of all kinds may start to make chirping or clicking notes a day or two before hatching, but in species that leave the nest quickly these noises serve a special function. Experiments with Bob-white and Old World Quail have shown that these notes speed the rate of hatching in slower chicks so that they will all hatch at once, even if incubation of individuals begins as much as two days apart.

Unlike precocial parents, birds that begin incubation after laying the first of several eggs continue incubating until the last-laid egg has hatched. Since the newly hatched chicks must be brooded much of the time, this delay does not inconvenience the parent still incubating, because its mate brings food for the chicks. An egg that fails to hatch on schedule may be incubated for several more days, until the urge to feed the young overpowers the urge to incubate, but if none of the eggs in the nest hatches, the parent may continue incubating long after the normal time span; many passerines will sit on eggs for twice the usual incubation period. A Ruffed Grouse sat on infertile eggs for 70 days, nearly three times longer than the incubation period, and an Anna's Hummingbird covered her eggs for 95 days.

Birds whose chicks remain in the nest for any length of time after hatching always remove the eggshells and membranes, which could attract predators by their strong odor and conspicuous appearance. This also avoids the danger that a piece of shell might cover an unhatched egg, making it impossible for the chick to break through. Most birds carry the shell away from the nest before dropping it; the fragments of shells you find on the ground in spring and summer are nest discards. Some birds eat the shell, which may have mineral value. Eggs that fail to hatch are not removed from the nest.

Growth and Development of Young

When the chick hatches, its appearance depends on how much growing it has done within the shell. Precocial chicks like the Killdeer overleaf, which will leave the nest a few hours or days after hatching, are covered with down that dries within hours of emergence from the shell. Their eyes are open, they are alert, and they respond to their parents' warning notes by crouching motionless even before they are old enough to recognize danger for themselves. Precocial birds include all the waterfowl, grouse, quail, pheasants, rails, cranes, plovers, and sandpipers. Most are capable of recognizing and securing food on their own as soon as they leave the nest; others, like the domestic chicken, depend on their mother to scratch the ground and point out edible items for the chicks to pick up.

The three eggs of the Herring Gull are laid and hatch one day apart. Here one chick is breaking through the shell, the second egg is pipped, and the last-laid egg's chick has not yet begun to work on the shell.

A Killdeer chick a few hours after hatching.

Altricial birds may be born with or without down. They all remain in or near the nest and depend on their parents to provide food. Gull and tern chicks are born covered with down and may wander short distances from the nest, but are entirely dependent on their parents, while herons, tubenoses, hawks, and owls, also born with a down covering, stay in the nest for at least a few weeks. All passerines, hummingbirds, cuckoos, pigeons, pelicans, cormorants, as well as hole-nesters like woodpeckers, kingfishers, and parrots, are born naked or with only sparse down on their feather tracts. Their eyes are closed and at first their small legs cannot support them when they rise to beg for food. Before they can see their parents, sightless young birds begin begging when they hear their parents call or feel the movement of something landing at the nest. Very

A precocial chick at hatching: the down of this American Coot will soon dry and fluff up, and in a few hours it will swim away from the nest.

A typical altricial chick like this Eastern Bluebird is born blind and naked, but with the instinct to raise its head to beg for food.

young birds will call for food if you gently shake the branch or bush holding their nest.

A prominent feature of young passerines is the bright color of their mouth linings. When begging, young passerines open their mouths as wide as possible, exposing a large area of bright red, yellow, or orange that helps the parent find the right place to put the food; this is especially useful to birds nesting where there is little light. The color of mouth linings is uniform in most families — finches, tanagers, and blackbirds have red linings; vireos, thrushes, wrens, nuthatches, swallows, and New World flycatchers have yellow or orange. The mouth lining of each species of African grassfinch has a distinctive pattern of dark markings. Similarly, in birds where the chick pecks the bill of the parent when it wants to be fed, the parent's bill has a distinctive mark indicating where to peck; the red dot on the lower mandible of the Herring Gull and other gulls serves this purpose.

All chicks, no matter how covered with down, need to be brooded; young birds cannot regulate their body temperature, so must be warmed periodically, and covered during cold spells, at night, and in storms. Altricial birds are brooded almost constantly in the first stages of nest life; brooding gradually decreases as they develop feathers and the ability to regulate body temperature. Young birds may continue to be brooded at night until just before they are ready to leave the nest. Birds whose nests receive no shade often stand over their young on bright or hot days.

Parents of altricial birds must devote most of their time to feeding the young. Except in those birds noted already, where one parent is entirely responsible for the young, feeding is usually done by both parents. During the first few days after hatching, the male may

A Blue-gray Gnatcatcher bringing food to its young; the nest, usually high in a tree, is typically bound by spiderwebs and decorated with lichens.

bring most of the food to the nest while the female broods the young; later both gather food.

The rate at which young are fed varies widely, even within one species, depending on age, time of day, weather, and type of food. In the first day or two after hatching, they may be fed very little, as they are still getting nourishment from the yolk-filled abdomen, but feeding rates then increase rapidly. Small insect eaters hunting close to the nest may bring food every two or three minutes: a pair of Barn Swallows was seen feeding its young once a minute. Larger birds searching farther from the nest feed their young less frequently, but with larger loads. Cedar Waxwings may visit the nest every twenty minutes, but with enough berries to feed every chick.

Raptors may not find food for the young more than once every few hours, but often it is a large carcass that can be left at the nest for the young to feed on when hungry. The young of raptors can also gorge themselves when food is available and then survive for hours or days without eating. White Pelicans bring large loads to the nest twice a day after extended fishing trips, and small petrels feed their single chick once a day; the larger tubenoses travel far over the open sea for food and visit the nest only once every several days.

Among birds that feed their young frequently throughout the day, the peaks of activity seem to be in the morning and in late afternoon and evening. Weather may change this if it affects food supply. An extreme example is the European Swift, which is unable to find flying insects during cold or rainy weather: during these periods of hours or days, the young become torpid and do not require food; they can survive up to 21 days without food, although 10 is the average. The nest period of the Swift therefore depends on the weather; in England it varies from 37 to 56 days.

Most young birds are fed the same types of food they will eat as adults, although in the early stages of development fruit and seed eaters often give their young a higher percentage of insects, which contain more proteins and other nutritional elements needed for growth. Cedar Waxwings, for example, are fed insects only for their

Brown Pelican feeding young.

first three days. Goldfinches, an exception, feed their young a pulp of predigested seeds, without insects, from the beginning.

Some birds that take larger prey make an effort to feed their young items of suitable size. Common Terns bring very small fish to their newly hatched young; the size of the fish increases as the chicks are able to swallow larger ones. Many raptors tear a carcass into small pieces that they give directly to young chicks; when the chicks are older, they tear the carcass apart themselves.

Food is brought to the nest in the beak, or regurgitated from the crop in the throat, or, by some raptors, carried in the talons. Most young receive the food by opening their mouths wide so the parent can place an item deep down the throat, but young herons grasp their parent's bill and shake it to make the adult drop fish into the nest or directly into the bill of the young, and all pelicaniformes (pelicans, cormorants, boobies, frigatebirds, etc.) except the tropicbirds open their bills, letting the young pick fish out of their throat.

Pigeons, primarily grain and fruit eaters, produce a unique food for their young. Called "pigeon's milk," it is a secretion containing vitamins, proteins, and fats, produced by both sexes by a sloughing off of cells lining the crop; the young switch to the adult diet after they leave the nest.

In nests with more than one young, the hungriest, noisiest, most energetic beggar is fed first. When that bird is temporarily satisfied, the next most energetic is fed. This system gives the advantage to the stronger chicks, or the eldest in asynchronous nests, but if the parents can find enough food, all will be fed adequately. If food is in short supply, at least some individuals will get enough — an even division might not give any chick all it needs.

Growth rates of young birds vary widely. It is advantageous for all chicks to mature rapidly, since they are usually more vulnerable to predators when in the nest or before they are able to recognize and avoid danger. Balanced against this is the ability of the parents or the precocial chick to find food: organisms cannot mature more quickly than they are nourished. The larger albatrosses, for example, feed their single chick only once every several days; the chick remains at the nest until 8 or 9 months old. As we have seen in the previous chapter, some hummingbirds can raise two or three broods in the time it takes other species to raise one.

Among passerines, growth and development are rapid — American Robins, for example, born weighing 5.5 grams, leave the nest 13

These young American Robins are almost ready to leave the nest.

days later weighing 56 grams. The eyes of most passerines open 3 to 5 days after hatching; in a few days they can see well and begin begging noisily for food at the sight of their parents. At this stage they also develop a sense of fear and will no longer let themselves be held docilely; if picked out of the nest young birds now react by freezing or by frantic calling. Juvenal feathers unfold from their sheaths about midway through nest life, so that when the young are ready to leave the nest, feathers cover the body, although wing and tail feathers have not yet grown to their full length. In the last days of nest life many birds begin stretching, preening, and exercising their wings. If startled, they may "explode" from the nest before they are really able to fly, temporarily avoiding one predator but, until they can fly, remaining easier victims to another; for this reason it is best not to disturb nests of birds not quite ready to fledge.

Hole-nesters mature more slowly than open-nesters, because their nest site is safer. They also leave the nest at a relatively later stage of development, when they can fly well, even though they have no room to practice flapping their wings while in the nest.

Larger birds of course take longer to grow; hawks and owls stay in the nest 6 or 8 weeks, eagles about 12 weeks.

In the tropics, where passerines have longer incubation periods, the time spent in the nest is also longer. It would seem that tropical parents have the time to gather more food for their young, especially since, as noted before, the brood size is smaller than in temperate regions; why their young grow more slowly is another mystery of tropical ecology. Tropical birds will not renest as quickly as temperate species if the first nest is destroyed, and spend comparatively longer caring for the young after they have left the nest than do temperate relatives. Since the entire nesting process takes longer for tropical birds, the longer period of postfledging care may be to insure that all the earlier effort is not wasted; tropical birds apparently produce more offspring by remaining with the young longer than by attempting to renest.

Most altricial birds practice some form of nest sanitation to eliminate sights and odors that could attract predators, to reduce the number of nest parasites, and to keep the young from fouling or matting plumage needed for warmth; the young of many altricial non-passerines like herons defecate at or over the edge of the nest and young passerines void their excreta in a fecal sac that can be easily removed by the parents. Usually the young defecate immediately after being fed, while the parent waits. Some birds then eat the fecal sac, which, because the young bird's digestive system is still inefficient, contains some nourishment. As the young grow older and absorb more of the nutritive value in their food, parents stop eating the feces. Other birds carry the sacs away from the nest, dropping them at a distance as they dropped the eggshells. These habits vary individually as well as between species. Some hole-nesters, including kingfishers and most woodpeckers, do not practice any nest sanitation, although the passerines like chickadees that later nest in old woodpecker holes always do. If a chick dies in the nest it is carried away only if small.

The relations among siblings in a brood are usually peaceful. Altricial young begging for food rarely attempt to take what has been

A Red-breasted Nuthatch removing a fecal sac from the nest.

A Lark Sparrow bringing food to young that have just left the nest. They will be fed several more days, until they reach adult size.

given to another chick, and the system of feeding the most eager beggar and of putting the food far down its throat reduces the possibility of theft. If parents bring a food item too large for one chick to handle, more than one sibling may try to get it; two young Common Terns may both pull at a fish which one has dropped, but neither directly attacks the other to get the fish. Gannets and boobies sometimes lay two eggs, but rarely can supply enough food to raise two young; in the Brown Booby, the elder chick usually evicts the younger one from the nest, so that it isn't fed at all and eventually dies. Cannibalism occurs in many raptors, most frequently among the larger eagles that produce only two young. The killing and eating of the younger chick by the elder is so regular a habit in the Lesser Spotted Eagle of eastern Europe that Czechoslovakian conservationists trying to increase the eagle population have a program of alternately removing and feeding each chick for a period of days until the younger one is large enough to resist attacks — the young eagles are apparently peaceable when eight weeks old. The common raptorial habit of leaving surplus food at the nest may serve to reduce cannibalism.

Among precocial birds that feed themselves but stay together for at least a few weeks, a rank system may develop. It has been found that in goslings final weight is correlated with the rank of each chick in the brood, not with its weight at hatching.

How the young leave the nest differs even within closely related types of birds. One factor is the location of the nest; many precocial species leave the nest a few hours after hatching, as soon as their down is dry, but Wood Ducks, whose young must jump from a tree hole, wait until the day after hatching. Similarly, the young of altricial birds that nest in holes or sites lacking nearby perches do not leave the nest until they can fly well, even though their cramped nest site has prevented them from so much as stretching their wings in practice. Whatever the circumstances of their departure, altricial young generally leave the nest unprompted by their parents. Small ground-nesting passerines usually leave 8 to 10 days after hatching, hopping after their parents before they can fly. Tree-nesting passerines leave at 10 to 14 days, when they can make at least short flights to perches near the nest, while small hole-nesters like the Black-capped Chickadee wait until 16 days old. All these birds continue being fed by their parents for several days outside the nest.

Post-nest parental responsibilities may be handled in several ways: the two parents may care for the chicks together, as occurs in most species, or each parent may take some of the young, or one parent may take complete charge. The last strategy is used by bluebirds and House Wrens; the male feeds the fledglings while the female begins a new clutch. In contrast, the European Swift flies well and is never fed after it leaves the nest site.

Seabirds have evolved a wide variety of nest leaving practices. The single young of some penguins and the Royal Tern leaves the nest when a few days old to form communal "nurseries" or "crèches." Here, large groups of young stand or walk through the colony in a loosely packed flock. If alarmed, young of a Royal Tern crèche rush into the water, returning to shore when the danger seems past. The young terns are not guarded by any particular adults, but there are always at least a few adults around the crèche, feeding their own chicks. Although a Royal Tern crèche may contain thousands of birds, parents and young recognize one another by their calls and the distinctive markings of each chick, so feeding is not indiscriminate as was previously believed. When able to fly, at three or four weeks, each young Royal Tern leaves the crèche with its parents, who tend the chick through the following winter, when its fishing abilities are sufficiently developed to allow independence.

Young murres and Razorbills are similarly tended by their parents after leaving the nest; when two or three weeks old and only one-third its parents' weight, the chick flutters down to the sea from its birthplace on a cliff ledge, accompanied by at least one parent, who will lead it out to sea and continue feeding it — an early departure from the crowded cliff ledge may reduce the possibility of large but clumsy chicks falling off.

A different strategy is used by some seabirds nesting in safer sites: the Common Puffin, a close relative of the murres and Razorbill, and the Manx Shearwater, which both nest in burrows, feed their chick for several weeks, until it weighs more than the adults, when it is deserted; the chick stays at the nest for 7 to 10 days, losing weight, and then flutters to the sea, where it cares for itself. Many of the albatrosses and larger shearwaters that feed their young at increasingly long intervals were thought to similarly abandon the chick. This has been verified for some of the migratory species; whether deserted or not, the young tubenose is not cared for in any way once it leaves the nest.

The Learning Process

Recent experiments and research with wild birds have shown that the learning process is more complicated than had been thought. While many behavior patterns are inherited and purely instinctive, other basic activities and skills must be learned. As we saw in Chapter 7, the development of vocalizations varies greatly from species to species in its responsiveness to experience.

Learning begins prior to hatching; before the eggs hatch, many birds give calls near the nest that are then recognized by the young as soon as they emerge from the egg. Experiments with Laughing Gulls revealed that chicks hatched in incubators where tapes of alarm and attraction calls had been played responded correctly after hatching, while those from silent incubators did not. Similarly, Common Murre chicks learn to recognize their parents' voices while hatching.

Immediately upon hatching, young birds that are led away from the nest by a parent must learn that the parent is the correct object to follow, and must not be diverted by others. Imprinting, as this learning process is called, is especially strong in ducks — newly hatched ducklings usually imprint on the first living object they see. Experiments have shown that this object must either move or emit

A nest of the Old World Little Egret, with the young beginning to climb on to neighboring branches, as a parent watches.

short, low, repetitious sounds; these requirements prevent a duckling from imprinting on an inanimate object. In hole-nesting species, like the Wood Duck, imprinting is by sound, since the ducklings must follow their mother's voice to get out of the nest. Imprinting always occurs within a few hours of hatching, and is nearly impossible to change thereafter. Pet ducks and geese that follow people around and that often reject other waterfowl as companions do so because they were imprinted on a person. Since imprinting plays a role later in mate selection, it is a learning process of crucial importance throughout a duck's life.

Recognition of the parents is of less immediate importance to most young birds, and in some species the parents learn to recognize their young first. The more opportunities there are for confusion, the earlier recognition occurs; Royal Terns, whose young are close together on separate nests even before they join the crèche, recognize their young from the time of hatching — this ability has been aided by the evolution of a wide variety of down plumages. Herring Gulls, with more widely spaced nests, take five days to recognize their young; this is about the stage when the young start straying from the nest. Recognition of either parents or young occurs even later in species with widely separated nests where, under normal circumstances, there is less chance of mistaken identity.

As they mature, altricial chicks become increasingly interested in the world around the nest. After they can see well, they become very excited at the arrival of a parent. Birds in spacious nests, like herons, may pick at the sticks of the nest and eventually start climbing onto the branches at the nest rim. Small birds may snap ineffectually at passing insects. When feathers have developed, birds preen, stretch, and, if there is space, may flap their wings.

Flight and feeding are the most important skills a bird must acquire, but, as we have seen, many birds can fly well without any practice and can locate and capture food on their own as soon as they leave the nest. Feeding ability is slower to develop in some fishing and hunting birds, but few cases of actual teaching by the parents have been recorded. Some evidence that feeding skill improves only after experience is that most raptors and fish eaters do not breed in their first year after hatching. Royal Terns eight months old and still periodically fed by their parents were found to be only half as efficient fishers; they do not nest until three years old. An interesting case of teaching was observed in a Belted Kingfisher that dropped fish that were beaten but alive into the water for its young to retrieve; in ten days the young were catching their own fish.

Helpers at the Nest

Assistance in caring for young birds by individuals other than the parents occurs occasionally. The helper may be a bird that has lost its own brood but still has the urge to feed or guard young. Sometimes young birds that have lost their parents are adopted by others. In the Common Eider, where the male does not help care for the young, females often join their broods together, and are sometimes assisted by non-breeding first-year birds. Very occasionally, confused birds may leave their own nest to help raise young of another species. In England, for example, a pair of Blue Tits built their nest in a box on top of which a pair of European Robins was already nesting. When the robin chicks hatched, the tits covered their own eggs with feathers and fed the robin nestlings. At first the two pairs fought a little, but then raised the young robins together. After the robins had fledged, the tits laid another set of eggs over their old ones and raised this brood (with no help from the robins).

Nest assistance on a more regular basis occurs in the bluebirds and Barn Swallow. Young of the first brood often help feed those of the second. In the Chimney Swift, single adults sometimes help other pairs with incubation, brooding, and feeding young. These unattached individuals may be first-year or old birds.

Truly communal rearing of young occurs only in sedentary species. The anis, which live in groups and jointly build a nest where several females deposit eggs and the entire group cares for the young, have already been mentioned. In most communal species, however, one female lays all the eggs. This is the case in the commonest North American communal nester, the Acorn Woodpecker, where a group of up to ten individuals jointly defends a territory. The groups seem to be composed of at least two older birds and some first-year birds. All the birds help one pair of adults by feeding and brooding young and by defending the nest.

Jays of different species show several stages in the evolution of communal nest care, with the more highly developed forms occurring in the tropics. Breeding pairs of Piñon Jays may also feed other fledglings, but first-year birds give no assistance. In Scrub Jays, first-year birds defend the territory and nest, and feed the young. The Mexican Jay (which also occurs in the southwestern United States) lives in flocks of as many as 20 birds that jointly defend a territory in which two adult pairs of the flock build separate nests; the pairs are each assisted by some of the non-breeders in the flock, which help bring nesting material and later feed the young. Flocks of other more tropical jays have an extended breeding season, with a second nest begun sometime after the first; in these species several individuals may take turns incubating, as well as feeding the incubator, guarding the nest, and feeding the young.

Communal nesting has been found in about 80 species of 32 families, and the list is still growing. The ecological features shared by most of these species may help explain the habit's evolution. Communal nesting takes place mainly in non-migratory species where young enter the population faster than mature birds die off. Their environment is constant and choice habitat may be limited. There is a long time between breeding seasons, so young of the previous season are competent to help, although perhaps not competent enough to raise a family of their own. The young gain valuable experience and the adults succeed in raising more young with assistance than alone. That most of the nest helpers are males probably indicates that females mature sooner and leave the flock to start their own, not that an uneven sex ratio in the population causes an excess of males.

Parasitism

Depositing eggs in the nests of other species to be incubated and, after hatching, fed is a reproductive strategy that has evolved several times. It occurs in ducks, cuckoos, honeyguides, grassfinches, and cowbirds. Except for the duck, all the brood parasites have evolved complex mechanisms for the toleration of their eggs and young in the nests of other species: some eject the eggs or young of their hosts and others have young that mimic the calls and/or appearance of their nestmates and live peacefully in the nest together.

Many ducks occasionally lay eggs in the nests of other ducks or coots, but the only exclusively parasitic species in the family is the Black-headed Duck of southern South America, which deposits its eggs indiscriminately in the nests of coots, other ducks, ibises, spoonbills, herons, storks, and gulls; all accept the extra egg. When the duckling hatches, its only need is to be brooded for two days, after which it swims off and feeds itself. It is never fed by its host.

Parasitic cuckoos occur on all continents except North America. Some, like the common European Cuckoo, have evolved eggs that perfectly match those of their hosts. To further insure success, the female Cuckoo removes an egg of the host's when she lays her own; this egg is removed before her own is deposited, eliminating the possibility of the Cuckoo removing her own mimetic egg. The Cuckoo parasitizes many species, but each individual lays only in the nest of the one species whose eggs are mimicked by hers. The Koel, an Asian cuckoo, has young whose voice and appearance also

HOSTS PARASITES

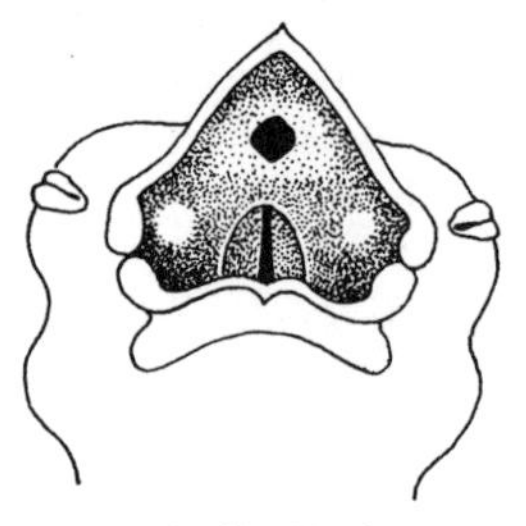

Melba Finch

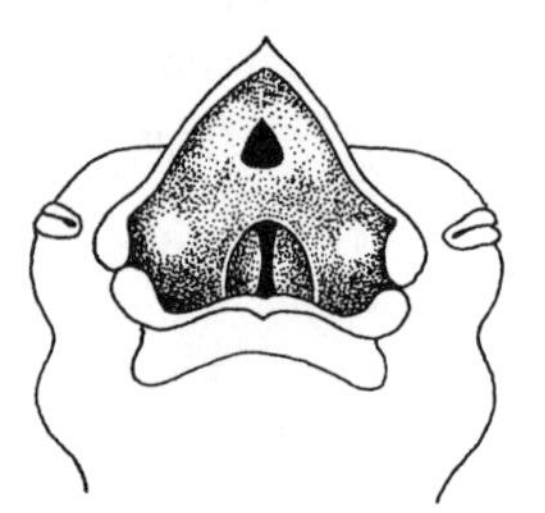

Paradise Widow-Bird

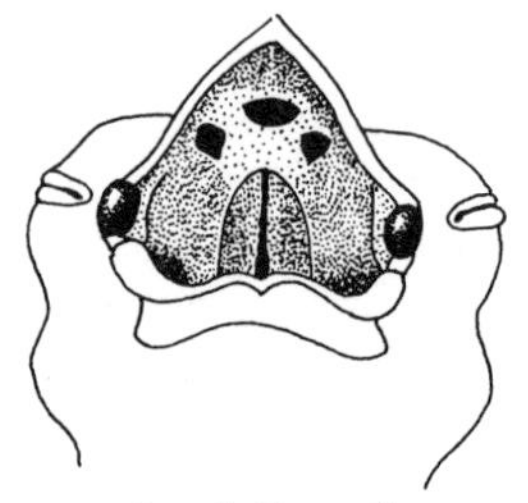

Purple Grenadier

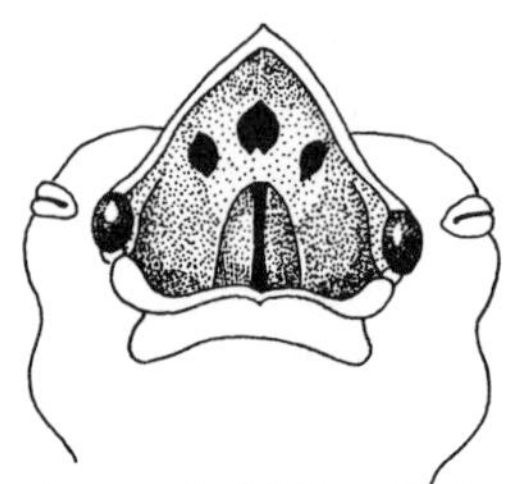

Straw-tailed Widow-Bird

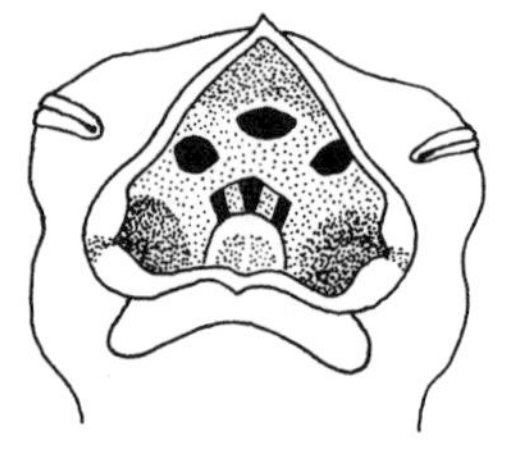

Jameson's Fire Finch

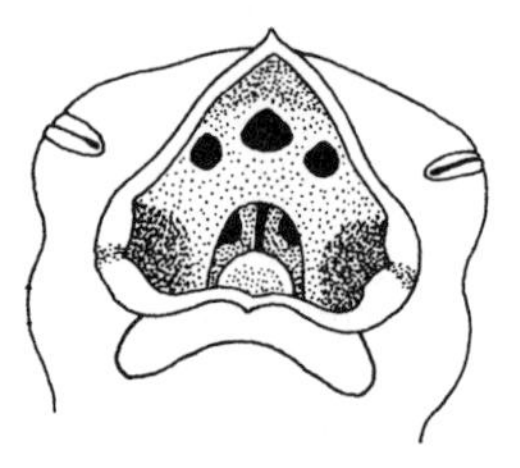

Purple Combassou

Mimicry of host mouth-linings by parasitic grassfinches (from "Mimicry in Parasitic Birds," by J. Nicolai. Copyright © October 1974 by SCIENTIFIC AMERICAN, INC. *All rights reserved.)*

closely mimic the young of their usual host, the Indian Crow. The young of some non-mimetic cuckoo species have a depression on their back that they use to shoulder the host young out of the nest.

In Africa, grassfinches lay their eggs in the nests of other finches. The host species have evolved complex, very species-specific patterns on the mouth linings. Parents will put food only into mouths with the right pattern, so each parasitic species has evolved a pattern matching its host's, as shown opposite. Mouth patterns probably first evolved in the host species as an anti-parasite device.

The honeyguides of Africa and Asia primarily parasitize barbets, woodpeckers, and other hole-nesters; the host, even if it detects an egg as foreign, cannot remove it from the hole. Young honeyguides have a powerful hook on the tip of their bill that they use to kill their nestmates, assuring themselves a monopoly on the food brought to the nest.

Cowbirds, in the New World, have not evolved any specialized mimicry; instead, the female cowbird often removes one of the host's eggs, and the young cowbird hatches sooner and grows faster than its nestmates. The nestling cowbird is often larger than its foster parents and crowds the other young out of the nest. The Brown-headed Cowbird of North America lays its eggs in the nests of many species and, like most parasitic species, is territorial, so that cowbird chicks do not usually compete with one another.

The several tropical cowbird species illustrate the different stages of parasitism: the Bay-winged Cowbird occasionally builds its own nest, but usually robs other birds of their nest or uses an old one to raise its own young; the Screaming Cowbird parasitizes only the Bay-winged; the Bronzed Cowbird usually parasitizes the related orioles; the Shiny and Brown-headed Cowbirds parasitize many species.

For a parasite to be successful, it must be less common than its hosts, assuring a continued supply of host nests. Parasites do not significantly alter the population of their host, and usually victimize only very common birds whose nests are easy to find. One exception to this may be the Kirtland's Warbler, which has only recently become a host of the Brown-headed Cowbird. Limited to a few hundred pairs at most, Kirtland's Warblers cannot afford to lose the number of nests currently being parasitized. A recently begun program removing cowbirds from the warbler's restricted breeding areas may help.

This Chestnut-sided Warbler feeds a Brown-headed Cowbird chick that is already as large as its foster parents and has pushed their young out of the nest.

The feeding ecology of brood parasites may explain the adaptive advantages of parasitism, although not how it actually developed. Many parasites eat items not well suited to young birds — some cuckoos specialize in hairy caterpillars, honeyguides eat wax, and grassfinches take grass seeds, so the young of these species may be better nourished on a diet supplied by less specialized feeders. The Brown-headed Cowbird was originally a mobile follower of migrating buffalo herds; were it tied to a particular location for the time needed to raise young, it could not maintain an association with the moving herds. Some other cowbirds are also roving feeders. (Conversely, it is possible that the parasitic habit is what enabled cowbirds to become roving feeders.) Since some normally nonparasitic birds, including the North American cuckoos, occasionally lay eggs in the nests of other individuals or other species, the parasitic habit may have developed from this originally aberrant behavior.

Many host species have evolved anti-parasite behavior: Blue Jays, Eastern Kingbirds, Cedar Waxwings, Gray Catbirds, Brown Thrashers, and American Robins are among those that puncture and eject Brown-headed Cowbird eggs; Yellow-breasted Chats desert the nest if they find a strange egg in it, and Yellow Warblers build a new layer over the egg and lay a new clutch, sometimes as many as five times if repeatedly victimized. The birds that react to cowbird eggs recognize them only as different from the others in the nest; given a set of four cowbird eggs and one of their own, they eject their own.

The variety and complexity found in breeding strategies, finely attuned to the pressures of each species' environment, indicate the difficulties involved in finding a place to nest and in raising young. In spite of all the precautions taken, many attempts at nesting are not successful. Mortality rates are higher for some species than for others, but all birds have the problems of infertile eggs, embryos that die at some stage of development, nest destruction by weather or predators, fluctuating food supplies, parasites, disease, and the possibility of one or both parents dying before the young are independent.

A nesting success rate of about 50 percent may be the average for passerines with open nests. Hole-nesters may do slightly better. Individuals that survive the nesting season must still face the challenges of other seasons, which may include two migrations, a severe winter, food shortages, and the usual predators. By the following

When a Yellow Warbler finds a cowbird egg in its nest, it covers all the eggs with a new layer and lays again; this nest has been parasitized twice, but nests with as many as five layers have been found.

breeding season, the population is approximately what it was the year before.

Birds have the ability to recover population losses caused by temporary or unnatural phenomena. The recovery of many species, like the Brown Pelican in California, since use of certain pesticides has been stopped, is extremely encouraging. Conversely, changes in the environment that temporarily make breeding conditions more favorable and allow more young to survive, as a spruce budworm outbreak does for warblers in northern forests, will not cause a long-term population increase; when conditions return to the norm, any surplus population will be unable to survive.

Continued study of the life histories of the many species still imperfectly known will give us a better understanding of how each bird functions in its environment. Long-term investigations are needed to explain population dynamics and mortality rates, particularly in long-lived and colonial birds, but even a single season spent watching one pair of birds attempt to reproduce will give you insights into the ecological pressures faced by all birds.

Chapter Ten

Migration

FEW ASPECTS of bird study interest as many people as does migration. Among the most exciting times for bird watchers are the "wave" days, when many migrants arrive together, or the discovery of a vagrant off its normal route. Beside the day-to-day excitement of following the migration, noting arrival and departure dates, high and low counts, or rarer species in your area, there is the deeper fascination of discovering how individual birds can return year after year to the same location, how they navigate over unfamiliar, sometimes featureless, land and sea, and how migration began in the first place.

To fully appreciate a migration season, keep a list of all the birds you see every time you go bird watching. Note the number of individuals of each species and, if possible, which age or sex arrives first. Patterns of arrival and departure will soon develop: some species may move through your area very quickly, some may be present for several weeks before departing to nest or winter elsewhere, while others may come to your area specifically for nesting or wintering. Consider which birds are apt to arrive early or late, their habitat requirements, and how they utilize what is available. If you visit a pond, woodland, or meadow regularly, you will see that certain locations or plants may be attractive to birds at one season and ignored in another — the warblers that fed on insects in the spring foliage of deciduous trees may find more to eat in evergreens during the fall. What birds may avoid your area entirely, for lack of suitable feeding or resting areas? What are the effects of weather? As the seasons pass, you will be able to compare your records from year to year, noting any changes in schedules or populations, developing an understanding of the patterns of migration.

For as long as birds have been observed, people have been aware that some birds migrate, but many of the fundamental questions about how and why they do it are still unanswered. References to migration occur in the Bible, such as

> Is it by thy wisdom that the hawk soareth
> And stretcheth her wings towards the south?

from Job (39:29) and in Greek literature as far back as the *Iliad*.

The migration of large birds like the hawk mentioned in the Bible is easily observed, but other birds seemed to disappear overnight, as in fact they do, and many strange theories were put forth for their disappearance each fall and reappearance in the spring. Winter habits of familiar birds like swallows, the European Robin, and the White Stork were the subject of controversy and misconception for thousands of years. Aristotle, a careful observer of other bird habits, thought the European Robin was transformed into the Redstart, another European thrush, every fall, and retransformed in the spring, and swallows retreated to mountain peaks in the fall to shed their feathers and hibernate through the winter. Until the middle of the sixteenth century, no one doubted that swallows hibernated, although many writers felt they did so in mud at the bottom of lakes rather than on mountain tops.

With the increased exploration of the world in the sixteenth century came reports of storks and swallows present in great numbers in Africa at just the time they were absent from Europe, and in 1535 an account of a spectacular hawk migration in Cuba was published. Reports such as these led people in Europe to conclude that some large birds like storks and hawks, as well as swallows living in regions close to Africa, were migratory, but the swallows of northern Europe were still believed to hibernate. In 1703 an anonymous essay published in England asserted that swallows flew to the moon, a trip taking sixty days, where they hibernated until spring. Even Linnaeus, the most influential naturalist of the eighteenth century, was sure swallows hibernated in mud. Not until the nineteenth century, when many more naturalists had explored tropical parts of the world, was the concept of migration universally accepted.

Migration is usually defined as a regular seasonal movement from one place to another, with a later return to the first place, usually on an annual cycle and involving a departure from and return to a breeding area. Birds are the best-known migrants, but certain insects (especially butterflies); fish such as salmon, eel, and herring; and mammals including bison, caribou, and whales migrate long distances.

The reason so many forms of life have evolved migration as part of their life cycle is that their environments cannot support them all year — this usually means there isn't enough food to go around. Since most migratory birds withdraw after breeding from areas that become colder, many people assumed that it was the cold itself that birds sought to escape; however, it now seems clear that birds leave colder areas only because the seasonal change in climate eliminates most of the food on which they depend. Even birds that spend only the warmest months of the year in North America and winter in the tropics can survive a cold and snowy winter if they have enough food; with the increasing popularity of bird feeders have come many new records of usually tropical-wintering birds, like the Northern Oriole, spending the winter dependent on a feeder in northern states. Even species that never leave the tropics can, if fed, survive a northern winter: an escaped Troupial, a South American relative of the Northern Oriole, lived for two years in New York City's Central Park; during the winter it was fed orange slices, and on extremely cold days it buried itself up to its neck in dead leaves — behavior no Troupial in South America is likely to have practiced.

Cold is not the only aspect of climate that reduces food supply and forces birds to leave a certain area: in the tropics, many birds move about between wet and dry seasons, and some birds that nest in the northern Sahara Desert in early spring move north to the Barbary Coast for the summer, leaving the desert which by then has become too hot and dry to produce an adequate food supply.

The physical demands of long distance migrations are great, and, as we have seen, other activities, like molt, that require major energy expenditures are usually timed not to coincide with migration. To assure themselves of adequate fuel, many birds feed intensively in one location for several days or weeks just prior to long flights, building up fat reserves; some small birds like Blackpoll Warblers that fly directly from New England to South America or Red-eyed Vireos about to fly across the Gulf of Mexico have fat reserves amounting to 50 percent of their total weight.

Where Birds Migrate

In the Northern Hemisphere most migration involves a northward movement in spring and a southward movement when nesting activity is completed. As we have seen, however, migration is a response to a decreasing food supply, and for some species the winter food requirements are not due south of nesting areas; thus, migrations take birds in several directions. The many ducks which breed on the lakes of central Canada and Alaska must leave because these are completely frozen in winter, but instead of moving south to the southern states or the Pacific, large numbers travel much more east than south to winter on the Atlantic coast from New England to Chesapeake Bay. A land bird with an almost entirely east-west migration is the Evening Grosbeak; Michigan breeders winter in eastern states of the same latitude.

Some birds nesting in mountain ranges such as the Appalachians, Rockies, or Sierras simply move to different elevations for the winter. This sort of altitudinal migration is practiced by Dark-eyed Juncos nesting in the Appalachians — they move into the adjacent valleys for the winter, while Canadian populations travel hundreds of miles south.

A few Northern Hemisphere birds are required by environmental circumstances to move north for the winter: the Capercaillie, a large European grouse, leaves the leafless hardwood forests where it nests in Central Europe for the coniferous forests farther north; there it feeds on spruce needles all winter.

In the Southern Hemisphere, where the seasons are reversed, migrations exist as well, and the majority of migratory species move north for the winter and south again to nest. A look at a map of the world will show that the Southern Hemisphere lacks large land masses near the polar region and that the land areas of Africa and South America taper off to the south; these geographical factors mean there is much less land available to birds in the areas where food will be most reduced by cold weather, forcing birds to migrate, than in the Northern Hemisphere, and, as a result, migration is much less common in the Southern Hemisphere.

The longest migrations made by Southern Hemisphere breeders are performed by petrels and shearwaters. The Wilson's Storm Petrel nests on islands off Antarctica and southern South America; every Southern Hemisphere autumn the birds leave their nesting areas, fly

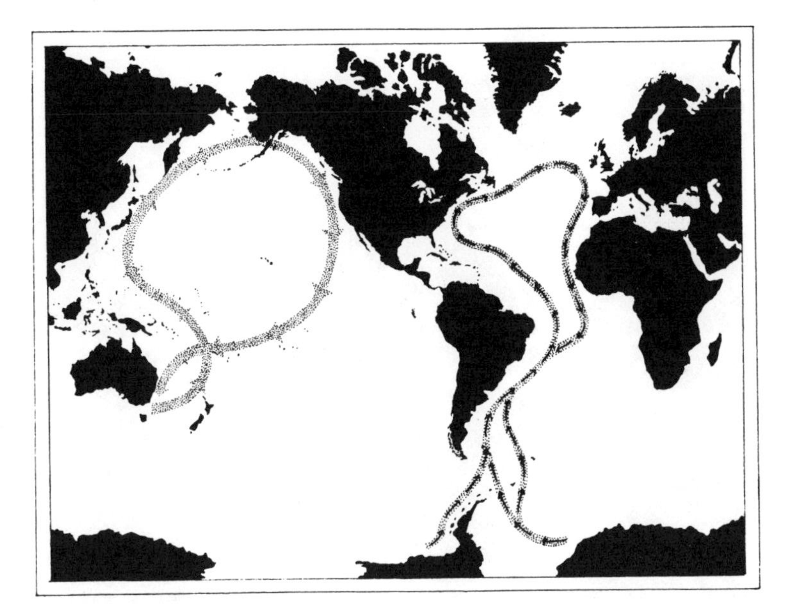

= *Short-tailed Shearwater*
= *Wilson's Storm Petrel*

The migration routes of two Southern Hemisphere breeders: the Short-tailed Shearwater and Wilson's Storm Petrel.

north through the Atlantic and into the Northern Hemisphere, where they are found throughout our summer in the Atlantic north to Labrador. The Greater and Sooty Shearwaters have similar migrations. All three species use a circular route, moving through the northwestern Atlantic in late spring and summer and returning to the Southern Hemisphere via the eastern Atlantic in fall. In the Pacific, the Short-tailed Shearwater, which breeds on islands off south Australia, has a similar route, moving clockwise around the north Pacific, north on the west side and south on the east side; the Sooty Shearwater also summers in the north Pacific, and other shearwaters that breed in Australian and New Zealand waters are seen off our Pacific coast regularly.

Very few other Southern Hemisphere breeders move into the Northern Hemisphere after nesting, but in every southern region there are species that move closer to the Equator. Almost every

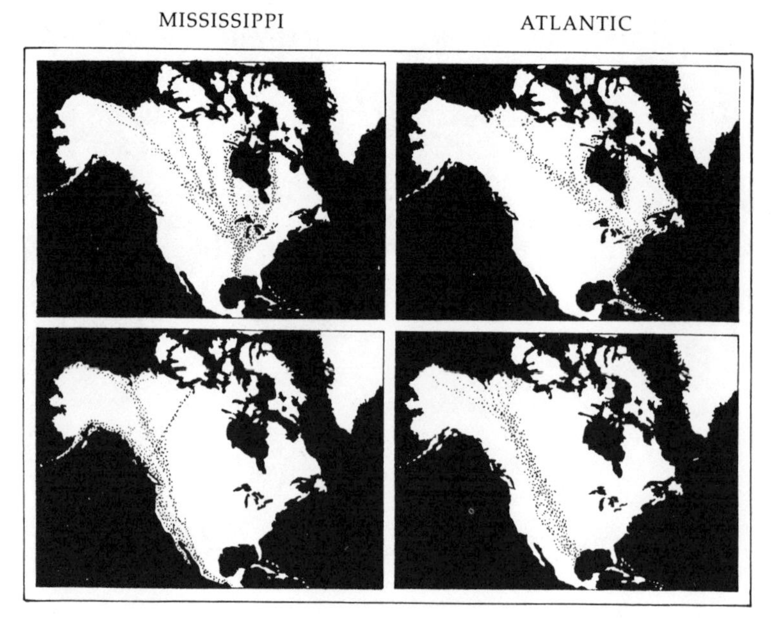

The major North American flyways.

group of Southern Hemisphere birds has some migratory species, including herons, waterfowl, shorebirds, terns, cuckoos, and passerines; only with the recent increase in study of Southern Hemisphere regions are we beginning to learn as much as is known about the more familiar migrations of the Northern Hemisphere.

In North America several hundred species are migratory, and there is no place on the continent where, sometime during the year, you will not see migrants. Some widespread birds can be found almost everywhere on migration while other species, including some abundant ones, travel over narrower routes. Certain routes are used by so many kinds of birds in both spring and fall that they are called flyways: North America has four major ones that follow prominent geographical features running north-south on the continent. It must be emphasized that while some flyways are named for specific features (Atlantic, Pacific, Mississippi) the term does not imply that birds are all flying along the shore or over the Mississippi River; the flyways are very broad paths and some scientists feel they are an inaccurate concept.

1. *Atlantic Flyway.* Besides those in eastern Canada and New England, birds from several northern regions including the Northwest Territories and Alaska move east and then down a broad belt from the Atlantic coast to the Appalachians to wintering areas in the southeastern states or beyond.
2. *Mississippi Flyway.* Some birds from many of the same Canadian regions that send birds to the Atlantic flyway use a route through central Canada, later following the Mississippi valley to the Gulf states.
3. *Central Flyway.* The most purely north-south route, east of the Rocky Mountains, ending on the Gulf coast of Texas.
4. *Pacific Flyway.* Most birds from Alaska and western Canada travel through the areas between the Pacific coast and the western edge of the Rockies.

Birds that winter beyond the southern states, in the West Indies, Central or South America, must travel to areas separated from North America by large bodies of water; some birds take routes avoiding the Atlantic Ocean and Gulf of Mexico, while others follow shorter but more hazardous routes over water. A look at the map will suggest at least six alternatives.

1. A few strong-flying shorebirds leave North America from Newfoundland, Nova Scotia, and Maine, flying south over the Atlantic Ocean to the Lesser Antilles and the northeastern coast of South America. No birds use the reverse route in spring.
2. Some birds that follow the Atlantic flyway to Florida go on via the Bahamas, Haiti, Puerto Rico, and the Lesser Antilles to South America.
3. Other birds leave Florida for Cuba and Jamaica, and fly across 400 miles of the Caribbean to South America.
4. A large number of the small birds that nest in the eastern United States and Canada, including the Ruby-throated Hummingbird, fly across the Gulf of Mexico from the Gulf coast to the Yucatan Peninsula and southern Mexico, an over-water crossing of 500 miles at the shortest point.
5. Birds that prefer to avoid long nonstop flights over water go around the Gulf coast to Mexico.
6. Western birds following the Central and Pacific flyways continue directly overland into Mexico.

Routes used by birds traveling beyond North America:

1. *Route of the American Golden Plover, Hudsonian Godwit, and some Blackpoll Warblers.*
2. *and 3. Used by the warblers, vireos, and others that winter on Caribbean islands or in northern South America.*
4. *Route of the Ruby-throated Hummingbird and many small passerines traveling to Central America.*
5. *and 6. These are used by birds that skirt the Gulf of Mexico or come south from the Rocky Mountain states.*
7. *Route of Pacific coast birds.*

The areas that provide suitable winter habitat for most birds are reachable by these major routes, and consequently they are used by the vast majority of North American migrants in both fall and spring, when the routes are followed in reverse. A few species, however, require winter environments that cannot be reached by any of these routes; their routes are adapted to fit their particular needs and capabilities.

The Arctic Tern, which nests throughout the far northern parts of the world and, in small numbers, on the Atlantic coast south to Massachusetts, winters far south in the Southern Hemisphere oceans, some individuals traveling 11,000 miles each way. The staple food of the Arctic Tern is the crustacea that abound in the cold waters of the Arctic and sub-Antarctic regions. The bird's long migration is a result of its dependence on food that is not available in sufficient quantity elsewhere, and its route is determined by the areas of ocean where food can be found between the breeding and wintering areas. Of course, unlike land birds traveling over water, the tern can move leisurely, feeding and resting as it travels. As shown on the map below, after nesting, Arctic Terns from North America west to Alaska travel across the Arctic and Atlantic Oceans to the coast of France, where they encounter other Arctic Terns from as far east as Siberia; the birds then move south to the west coast of Africa, where some recross the Atlantic and continue south off the coast of South America while others continue off the African coast before scattering over the South Atlantic and southern portions of the Indian Ocean as

The southbound migration routes of the Arctic Tern; return routes are less clearly known, but probably approximate these in reverse, avoiding areas of warm water.

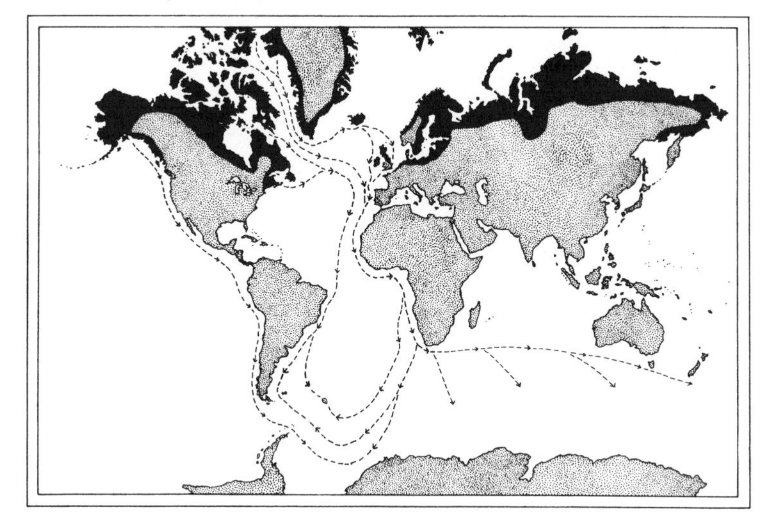

■ = *Breeding range*

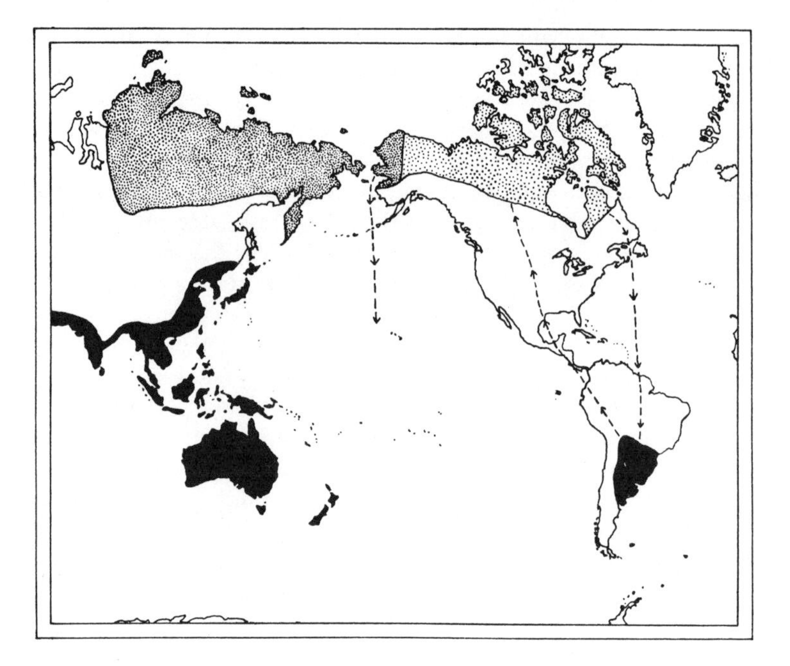

Atlantic Golden Plover breeding range
Pacific Golden Plover breeding range
Winter ranges

Breeding range and wintering areas of the American (or Lesser) Golden Plover. The boundary between the eastern and western populations in Alaska is not clearly known. The Siberian and western Alaska birds winter on all the islands of the Pacific and through much of Asia and the Australian region, while most of the North American birds follow the narrow paths shown to and from a restricted part of South America.

far east as New Zealand. This route avoids warm seas such as the western Atlantic where the preferred food of Arctic Terns is not available. Arctic Terns from Alaska move south off the coasts of North and South America and eventually encounter birds from the Atlantic. Return routes are less clearly known, but most eastern North American birds probably make the Atlantic crossings again to avoid areas of warm water. One consequence of this migration between polar regions is that the Arctic Tern spends more hours in daylight than any other bird, including about eight months each year in 24-hour days of it.

Another bird with distinctive migration routes is the American Golden Plover. Like the Arctic Tern, it breeds in the tundra and pond region of the far north, but unlike a tern that can spend months away from land, a plover must find a somewhat similar winter habitat; for most of the American Golden Plover population this is the pampas of southeastern South America. In fall there are few places between these regions that this species could utilize, so its migration is swift and direct. After nesting, the eastern population assembles in Labrador, Newfoundland, and Nova Scotia, where the plovers put on the fat they need for a long flight by feeding on the abundant crowberry. From there they fly nonstop over the Atlantic to the Lesser Antilles and the northern coast of South America, at least 2400 miles; many do not stop there, but continue directly over the jungles of the Amazon Basin to the pampas. In spring the birds take a different route, going northwest across South and Central America and the Gulf of Mexico to the grasslands of south Texas, where they stop to feed before continuing up the Mississippi Valley and Canadian plains to their breeding grounds. Some young birds, which leave the breeding grounds after the adults, use the spring route in both directions, perhaps because it is less hazardous than the long flight over water. The Pacific population uses an entirely different area in winter, flying south over the Pacific and scattering over all the islands from Hawaii to New Zealand, returning by the same route.

Few small birds perform migrations as arduous as these, but many do go as far as South America. The Bobolink's winter range overlaps the American Golden Plover's in Argentina and Bolivia: from the northern states and Canada, most travel south to Florida, then over Cuba, Jamaica, and across the Caribbean and, like the plover, pass over the jungles for the swampy regions farther south, while some Bobolinks fly across the Gulf of Mexico to Yucatan, then through Central America to the same destination; the same routes are used in spring. The Cliff Swallow winters in the same area, but avoids flights over water, flying around the Gulf of Mexico instead.

We have seen that some birds travel enormous distances between nesting and wintering areas. How far does the average bird travel? As a general principle one could say, "No farther than it has to." The American Golden Plover travels a long way, but knowing that it feeds primarily on insects in fields of short grass, can you find a suitable wintering area closer than the pampas? There are no others in Central or South America, and the cold eliminates most insects from the North American habitat that the plover uses on its way north in spring.

Likewise, the Arctic Tern goes to the areas where its food source, crustacea, is most easily available. The Arctic Tern also takes fish, especially when nesting, so one could ask why it doesn't continue to feed on fish, as other terns do. The answer may be that to join the several other species of terns fishing in the Caribbean and other warm areas every winter would create too much competition for a limited resource — the Arctic Tern may assure its own survival by traveling to an area with less competition for a different food source.

While these are examples of extreme migrations, they demonstrate that birds travel as far as they must to find an adequate food supply free of excessive competition. Wintering birds must compete for a limited supply of food with members of their own species, other migrants, and the native birds, often in a much smaller geographical area than they inhabit while nesting — the area of Central America and northern South America is considerably less than that of North America, yet it supports the majority of North America's insectivorous birds every winter; evidently, these regions produce enough food for most of the North American flycatchers, vireos, and warblers, etc., since few go very far south in South America.

Among hardier birds that remain in North America, ducks may go no farther south than the first available open water, while robins, blackbirds, and other ground probers move only as far south as the ground is usually soft enough to be probed, and seed eaters go only as far as sufficient seeds are left uncovered by snow. This does not mean that the entire population of any species is concentrated in the winter feeding area closest to the nesting area — the population of each of these species is too large for this to occur without severe competition, so the bulk of each species' wintering population is spread beyond the minimum distance between suitable nesting and wintering areas.

How Birds Migrate

The manner in which each species migrates reflects its needs and capabilities as closely as does its route and destination. The vast majority of birds fly, of course, but other species will swim or walk to their destination: several species of penguin swim north every year after breeding on Antarctic or sub-Antarctic islands, and even some strong fliers such as grebes, brant geese, eider ducks, and murres swim considerable distances. In North America the Common Turkey used to migrate great distances on foot as well as by flight; with the clearing of the eastern forests this is no longer possible.

The mystery of migration has always been heightened by the fact that many birds are never seen actually migrating. Birds that migrate at night baffled earlier observers, who could not account for their sudden appearances or disappearances, and today, with sophisticated scientific equipment, night migrants are still very difficult to study. Long flights pose hazards at any time, but for some species certain hours offer advantages which outweigh the dangers involved; others migrate during both light and dark hours, and many of the birds that usually fly during one period occasionally make such long flights that light or darkness overtakes them where they cannot stop, such as over the Gulf of Mexico.

What are some of the advantages of day and night travel, and to which birds might it make no difference? The majority of night migrants are the small land birds that travel great distances to the tropics and are ordinarily considered weak fliers; by flying at night they avoid being visible to hawks, gulls, and other birds that could easily catch them. Another advantage of night flight is that it allows all the daylight hours for feeding and rest; small birds burn up energy so quickly that after a night spent flying they must feed again, even before resting. Since these birds cannot fly as fast as some day migrants, but often have farther to travel, if they flew only in daylight in addition to feeding and resting, their migrations would take much longer, perhaps more time than the seasons allow, so for them flying at night is a much more efficient system.

Night migrants cannot usually be seen, but they can often be heard: many of the small birds periodically chirp; on a quiet night in April or May, or late August, September, or October, the sky may be filled with call notes that can be identified as to family, if not to species. The calls must help birds avoid collisions and indicate the number of birds nearby, although they cannot communicate any navigational information. On nights when the moon is full, birds can sometimes be seen traveling across it — try focusing on the moon with binoculars or a telescope; some larger species, like geese, can be identified by their profile.

Daytime migrants include the small birds that can feed as they fly, such as swifts and swallows, which catch aerial insects while traveling, combining two activities that are separate for most birds; moreover, they are fast enough fliers that they needn't make special efforts to avoid predators, as some small birds must. Other small daytime migrants include various finches and blackbirds, somewhat protected by flying in flocks and not forced to budget their time as tightly as the night migrants, since they are not traveling so far.

Many of the larger birds, such as hawks and herons, which often utilize thermals to keep them aloft with little or no expenditure of their own energy, must migrate during the day, because thermals exist only in the daytime and over land. To utilize these, some birds must increase their travel distance: the White Stork of Europe, for example, has a gliding flight that takes advantage of rising and falling air currents; it avoids flying over bodies of water like the Mediterranean, making long detours west to Spain to cross the Mediterranean at Gibraltar, or east to Turkey and Israel to avoid it entirely, before going south into Africa, while the European Crane, a bird of approximately the same size, flaps rather than glides, and is independent of air currents, taking a much shorter route across the Mediterranean at its widest point.

Still other birds, including ducks, geese, shorebirds, terns, and seabirds, travel both in the day and at night. They are strong fliers; some, like the American Golden Plover, have routes that require continuous travel over periods longer than those of light or darkness, and others, like the petrels and shearwaters that travel across the oceans, live in environments with hazards and benefits equal during day and night.

Weather conditions play a role in a bird's decision to migrate on any particular day or night. The weather factors influencing this decision are complicated, but obviously it is harder for a bird to travel in a violent rainstorm, in dense fog obscuring landmarks and hazards, or when strong winds might blow it off course; birds are found migrating in all these conditions, especially when they develop after the bird has begun flying, but birds usually try to take advantage of weather conditions that make flight easier. Most spring migrants move north with a warm front and seem to prefer a slight tail wind; fall migrants move south with a cold front and a north or northwest wind. Some larger birds fly most easily with a head wind, gaining support as a kite does — at Hawk Mountain Sanctuary in Pennsylvania, where large numbers of hawks pass every fall, taking advantage of the thermals created by the long series of mountain slopes, the greatest flights are seen when sunny local conditions combine with northeast or northwest winds and a depression moving east from the Great Lakes.

Birds travel at the heights and speeds that suit their requirements. Studies with radio transmitters attached to birds, with radar, and from airplanes indicate that most of the small night migrants fly below 5000 feet, and the majority between 2000 and 3000, but flocks of shorebirds are regularly detected by radar at 20,000 feet. The shorebirds are traveling great distances without stopping and may find favorable winds at that altitude. Sea ducks such as eiders and scoters fly just over the surface of the water, but rise higher over land, while land birds often fly higher over the sea. Flocks of blackbirds, jays, and finches sometimes fly just over the trees, while hawks riding thermals rise several thousand feet, beyond human vision. The highest altitude record is probably held by the Bar-headed Goose, which regularly flies at 30,000 feet to pass over the highest Himalayan peaks.

The rates at which birds travel is highly variable: some travel faster or longer during certain stages of migration, while many night migrants do not travel the entire night, and most spend a few days feeding and resting between flights. The migration of most small birds between the tropics and North America takes several weeks or months, but an altitudinal migration may be accomplished in a day. The Blackpoll Warbler, which flies from South America over the Caribbean to Cuba and Florida before spreading over North America, averages about 30 miles per day (but more per flight, since it does not fly every night) over the United States, but accelerates to 200 miles per day as it approaches its sub-Arctic breeding grounds; perhaps if its average speed were faster it would arrive north too early. In spring, many birds have to travel at a pace that is not faster than the northward progress of spring itself; sometimes insectivorous birds traveling north encounter a late cold snap, and may perish by the thousands if the weather kills their insect prey or delays its hatching.

Birds making long flights usually move at greater speeds: American Golden Plovers flying from Nova Scotia to northern South America, for example, usually make the 2400-mile trip in 48 hours, for an average speed of 50 miles per hour, and a marked Peregrine Falcon that escaped near Paris was recovered 24 hours later at Malta, 1350 miles away. Snow Geese make long nonstop flights between their

Arctic breeding grounds and wintering areas on the Gulf coast; some flights are more than 2000 miles, but the actual speeds are unknown.

Data from banding stations have been extremely valuable in plotting the speeds and courses of individual birds and have also revealed migration patterns that could not be discovered except by examining large numbers of birds in the hand. It is now well known that different populations of some species winter in different areas; female and immature Dark-eyed Juncos, for example, travel farther south than adult males, and immature Herring Gulls move less far south each year as they approach maturity. Banding data have also shown that different parts of the population may use different routes or schedules to reach the same destination. One such example of differential migration comes from a comparison of captures at coastal and inland banding stations: at stations on the Atlantic coastline and offshore islands, a very high percentage of fall passerine migrants are immature birds, while at inland stations the number of adult migrants may approach 50 percent, a more accurate part of the total population. It has been suggested that the more experienced adult birds take a slightly inland route, to avoid the dangers of being blown out to sea, while the immature birds, making the trip for the first time, fail to compensate for winds that drive them seaward.

Another aspect of differential migration involves adults and immatures migrating at different times in fall: at the Carnegie Museum's banding station east of Pittsburgh, it was found that the peak of adult movement for Red-eyed Vireos and White-throated Sparrows preceded that of immatures by ten days, and adult Dark-eyed Juncos preceded immatures by two weeks, but for Nashville and Magnolia Warblers, the bulk of immatures preceded adults by ten days; no explanation for this has yet been found.

Navigation

The most amazing and mysterious aspect of bird migration is navigation. How can birds traveling long distances, often in the dark, keep on a regular course and, in some cases, find their way back to exactly the same breeding or wintering spot used in previous years? What about the immature birds, often traveling alone, on their first trips? As we have seen, in some species the inexperienced birds fail to avoid the riskier route along the coast, and most of the out-of-place vagrants found in fall throughout the country, but particularly on the coasts, are immatures — disoriented western birds that have wandered to the Atlantic and Gulf coasts and eastern birds on the Pacific coast — but obviously the bulk of the first-time migrants must reach their proper destinations, or the species wouldn't survive.

Our appreciation of the navigational ability of birds has been considerably increased by experiments that have taken birds away from areas they normally traverse on migration, in which they could be expected to orient successfully. The most famous of these experiments were performed by Lockley on Manx Shearwaters nesting on the island of Skokholm off Wales. Shearwaters released in Venice and Switzerland were back in their nest burrows in fourteen days; whether the shearwaters took the most direct route, by flying overland (something shearwaters never ordinarily do) or returned

Homing experiments with Manx Shearwaters from the island of Skokholm, off Wales. Farthest release point was Boston, Mass., 3050 miles away. (*Reproduced with the permission of Pan Books Ltd. from* ANIMAL NAVIGATION *by Ronald Lockley.*)

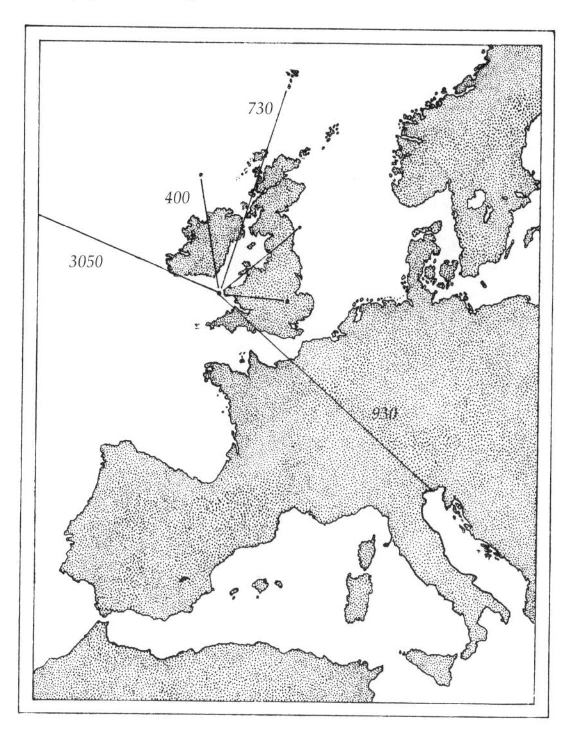

through the Mediterranean, the trips were over areas previously untraveled by Manx Shearwaters. In another experiment, a Manx Shearwater was released at Boston, 3050 miles from home, in another area unknown to the species; this bird was back in twelve and a half days. Similar experiments have demonstrated that several other seabirds, which normally travel far over the seas, have excellent navigational abilities, but many small land birds do as well — wintering White-crowned and Golden-crowned Sparrows returned to California from release points in Louisiana and Maryland.

Birds use several things to guide them, and most birds are capable of using more than one navigational system. The most obvious are geographical landmarks. Most birds that migrate at all move about to a certain extent within their breeding area and could recognize it from this experience when they returned the following year. Leaving the breeding area, day-flying birds could continue to follow geographical features such as river valleys, mountain ranges, coastlines, etc., most of which run, in North America, in the same directions that most migrants travel. Night migrants are of course active during the day, and may recognize landmarks as they near their destination.

By a series of experiments with Starlings in the 1940s and 1950s, the German ornithologist Gustav Kramer demonstrated that birds use the sun for navigation. Kramer observed first the behavior of Starlings kept in cages where they could see only the sun; in autumn the birds displayed "migratory restlessness," hopping and beating against the side of the cage, with most of the movement in the southwest side of the cage, the direction in which the Starlings would be migrating. In spring, the Starlings' restlessness was directed at the northeast side of the cage. By using mirrors, Kramer showed that the Starlings needed only to see the light, not the sun itself, to orient correctly — this is an important ability for birds, since the sun is often obscured by clouds. In experiments with artificial cloud effects, the birds became confused, but were able to reorient instantly whenever a short flash of light came through. When Kramer used mirrors to deflect the sunlight 90 degrees, the Starlings reoriented in terms of the light's new, unnatural direction, but other experiments showed they had an internal clock that compensated for the sun's normal movement through the sky.

Night migrants can use the position of the sun at dusk to get an initial direction, but they must use other means to maintain a steady course. Experiments by Stephen Emlen of Cornell University have shown how Indigo Buntings, and possibly other birds, use the stars. Emlen placed the buntings in individual cages made of a white blotting paper funnel mounted over an ink pad, so that every time the birds hopped on the blotting paper their feet left a mark; by counting the number of footprints in each part of the paper, he could determine the direction in which most movement took place. In the first experiments, out of doors, Emlen found that in autumn Indigo Buntings exhibiting migratory restlessness hopped mainly in the south part of their cages; under partially overcast conditions most continued to orient correctly. Brought into a planetarium where they were shown simulated fall skies, the buntings oriented correctly, but when the skies were reversed 180 degrees, the buntings similarly reversed their direction, remaining consistent with the stars.

By reducing the number of stars shown in the planetarium, Emlen learned which are important for navigation: he found that in both spring and fall the entire southern half of the sky could be eliminated without affecting the buntings' navigational abilities; removal of the Milky Way, Big Dipper, and North Star had little effect, but when all the northern stars were eliminated, even with the southern stars visible, orientation became random. If all the stars were eliminated, replaced by a diffuse light, orientation further deteriorated, and most birds ceased activity, which suggests that sight of the stars may, in the proper season, itself stimulate migration.

Ability to use the stars for navigation apparently varies among individual birds and between species and families, since among Emlen's buntings, different birds seemed to orient more accurately when shown different sets of stars, although they always oriented according to the actual season whether they were shown a spring or fall sky. In similar experiments, Old World warblers oriented according to the season of the sky they were shown; because migration has developed independently in different groups of birds, we might expect different uses or interpretations of the stars and other environmental stimuli.

Since birds sometimes migrate when weather conditions prevent them from receiving clues from either the sun or the stars, they must have other indicators also. In these cases wind direction may be an aid, especially at times when clouds obscure celestial features that were visible when the birds took off, but the wind continues from the same direction. The sounds of waves on shores, streams, wind through the trees, or animal life on the ground may be used by birds

— the writings of early balloonists, who heard such sounds while traveling quietly at great altitudes, have been used to show that these sounds could also be heard by birds; sounds could certainly tell them the habitat over which they are flying, and the sounds of waves breaking on a beach would be a valuable indicator to birds approaching land or trying not to leave it.

In addition to knowing how to reach their destination, birds must also know when to travel. A too-early spring departure may make birds arrive on their breeding grounds before enough food is available, and starting too late may mean missing the best season for beginning the nesting cycle. Similarly a too-early departure in fall may bring birds to their wintering areas at an inappropriate time, and by waiting too long they may be overtaken by the effects of cold weather.

Since so many birds arrive in and depart from their nesting areas at the same time every year, irrespective of day-to-day weather conditions, the sense of time seems to be independent of changeable environmental circumstances. For many birds, the initial migratory stimulus may be changes in day length; laboratory experiments have shown that some birds exposed to artificial changes in day length will assume the physiological state appropriate to the season the lights reproduce: thus, birds given increased amounts of light will develop the glandular characteristics of a bird ready to breed, and, if a nocturnal migrant, will hop and beat mainly against the northern wall of its cage in a state of migratory restlessness after dark, whether it is actually spring or not. However, it is not known whether the glands that have been stimulated by increased day length, in experiments or real situations, actually stimulate migratory urges or simply occur at the same time. If changes in day length indeed stimulate migrants, for those Northern Hemisphere birds wintering south of the Equator the stimulus must be decreasing day length in both spring and fall. Of course, this theory does not account for the development of migratory urges in the birds that winter in the equatorial zone of fairly constant day length; this is a subject on which much work remains to be done.

The Origin of Migration

A topic which will always remain theoretical is the origin of the migratory habit itself, and, as with so many habits of adaptive significance, migration may have evolved in several ways. One theory suggests that migration was made necessary by the first of the several periods of glaciation in the Pleistocene a million years ago; before then the Northern Hemisphere had a mild climate without cold seasons forcing birds to move or perish. Each of the ice ages forced birds to move south; as the ice sheets gradually retreated, the birds moved north again from the crowded tropics, and the development of regular warm and cold seasons as we now know them made migration a hereditary annual habit. The problem with this theory is that most of the strong-flying species of today existed at the end of the Tertiary period, millions of years before the first glaciation — perhaps sufficient seasonal variation existed even then in some parts of the world to stimulate migration. The most recent glaciations are important, however, to our understanding of the routes migrants take today, since glaciation has much to do with the shapes of current land masses.

Another theory suggests that at some time in the past the equatorial region became too crowded with birds, and some species moved north and south to nest in areas freer of competition, returning to their ancestral home after breeding. This theory works particularly well for families like the hummingbirds, flycatchers, orioles, and tanagers that exist in tremendous variety in the New World tropics and have fewer temperate zone representatives. Of the approximately 200 tanager species, for example, only three move far into North America, and these return to the tropics to winter; the ratios are similar for other families that probably originated in the tropics, but this theory cannot explain migration for the many families of birds that originated elsewhere.

So there are still many unanswered questions about bird migration. Some of the answers may come from more observation in the field, new banding data, laboratory tests, or new ways of looking at the facts we already know. To better understand the patterns of migration in your own area, get a weather report each night before you go bird watching in spring and fall and predict the kind of movement you will see the next day; in the field, you will see if your predictions are correct, and you may develop the same "judgment" a bird must use before it decides if it will migrate. Exercises such as this will give you an idea of the complexities about migration that ornithologists are trying to unravel.

Chapter Eleven

Winter Habits

EACH CHANGE of the seasons alters the environment in which a bird must find food and avoid dangers. Winter is generally the most challenging season, with severe stresses not encountered at other times; most birds change their behavior and many change their location to overcome or avoid these stresses. In the tropics as well, seasonal changes force birds to modify their behavior wherever wet and dry periods regularly alter the environment. The tropical bird's life is also affected by the climatic changes elsewhere that cause migrant and wintering birds of temperate regions to seasonally invade its habitat, where they may search for the same food or attract predators.

For birds of higher latitudes in both Northern and Southern Hemispheres, the most significant environmental changes that winter imposes are the reduced or altered food supply, the shortened day for finding food, and the colder temperatures that make the bird use more energy to keep warm. Some birds can survive these pressures in the environment where they were raised, and so live there permanently; the others move elsewhere. The migration strategy is obviously the only answer for the many birds whose food is eliminated by cold, but these birds must overcome the hazards of traveling and of living in different, at first unfamiliar, environments with new competitors. A few species, rather than migrating to an entirely different locality, simply disperse over the area near their breeding sites. Many Gannets of the British Isles colonies disperse over the sea at the same latitude, while the entire North American population, nesting on six Canadian coastal islands, migrates south, mostly to waters from Virginia to the Gulf of Mexico. The Herring Gull uses all three methods; adults may stay near their breeding colony, even roosting on it at night, or may disperse along the coast, while most of the younger birds migrate south. Individual Herring Gulls are apt to move a shorter distance south in each of their four sub-adult winters, perhaps indicating that survival is easier farther south for less experienced birds, or that, by dominating the better wintering localities near the breeding range, adults force the younger individuals to move away. Similarly, male Dark-eyed Juncos winter north of most females and immatures.

The change in day length is probably the most challenging factor for sedentary birds of the far north. Near Fairbanks, Alaska, for example, the day becomes as short as 3½ hours in midwinter, and in these few hours birds must find enough food to keep body functions going while they roost the other 20½ hours in darkness. The mean winter temperature at Fairbanks is approximately –13° F, and the temperature may drop to –58°. A study of Black-capped Chickadees there found that in winter, bright, sunny days with more minutes of light for foraging were often colder than the overcast days that had less feeding time. However, the bright days were so cold that the chickadees stopped feeding and went to roost sooner; apparently in extremely cold weather so much more energy is spent in foraging than roosting that it is more efficient for the birds to move less. This contrasts with studies of other birds farther south that roost earlier on warmer, overcast days than on bright, colder days, when they fed as late as possible.

Changes in precipitation also alter the habitat of sedentary birds. Most ground feeders must leave areas where the ground is thickly covered by snow, but the Rock Ptarmigan uses its heavily feathered, sharp-clawed feet to dig through the snow to the twigs and mosses on which it feeds. When the snow is too deep for digging, Rock Ptarmigans are attracted to bare, windswept areas. (Eskimos catch them by setting traps around an area they have darkened to make it look free of snow.)

The winter challenges to migrants are even more diverse. Aside from the dangers of migration and the stopovers in areas that may be different from the breeding area, wintering habitats usually differ in food items, vegetation, terrain, or predators. Southbound warblers of Canadian spruce forests, for example, must traverse several zones dominated by various broad-leaved trees, each with a different foliage structure, containing different fruit and insect food, where new predators using distinctive hunting techniques lurk in unfamiliar habitat.

The size of the winter range varies; in some widespread birds it is a large area that may overlap the breeding range, while others are much more restricted: the Cape May Warbler breeds in northern New England and in Canada from New Brunswick to south Mackenzie and Manitoba, a huge area, but winters in the West Indies, chiefly in the Greater Antilles and Bahamas — in winter, Cape May Warblers can evidently compete with each other in a much smaller space than while nesting. The Golden-cheeked Warbler, in contrast, breeds in dwarf forests of cedar and oak mainly in the Edwards Plateau of south-central Texas, but winters from southern Mexico to Nicaragua. The blue form of the Snow Goose winters primarily on the Gulf Coast, while various populations of the white form winter on the Atlantic, Gulf, and Pacific coasts, the Central Valley of California, and the Pacific coast of Asia south to Japan.

Movement to a different locality for the winter frequently involves a change in diet. Even raptors and insect eaters that take the same type of prey item year round may be consuming different species. Other birds make a more deliberate switch: many of the insect-eating passerines change to a diet of fruit and nectar when wintering in the tropics, reducing competition for a limited insect supply. Other tropical winterers change their foraging habitat — the Chestnut-sided Warbler breeds in open, bushy areas, but apparently feeds in the tropical forest canopy. The Blackburnian Warbler breeds high in trees; in winter it often forages in the understory and near the ground. The Brant eats a great variety of land and water plants in the Arctic tundra where it breeds, but feeds exclusively on a few aquatic plants like eelgrass and sea lettuce in the shallow coastal bays where it winters.

Winter also induces changes in the relations of individuals of the same species. Some that tolerated the presence of family members after they had become independent drive them away in winter. Others that bred solitarily now join flocks of one or more species. Birds that feed or roost in flocks are ecologically similar to colonial nesters in that they search for food that is irregularly distributed over a wide area. These flocks may number from a few tens or hundreds to the several millions that sometimes occur in mixed "blackbird" flocks of Brown-headed Cowbirds, Red-winged Blackbirds, Common Grackles, and Starlings. The larger flocks usually maintain the same roost throughout the winter and often year after year, while smaller flocks may wander. The roosts of large flocks are usually very conspicuous, but are often located in marshes, town squares, or other sites with a minimum of natural predators. The roost's large size further reduces the danger to each individual in it; small flocks, like a band of a dozen chickadees that feed together in the woods, always disperse to roost, since a group of this size is more vulnerable than either a single individual or a large flock. In winter Green-winged Teal gather by day in flocks on large bodies of water where they can most easily spot and avoid predators, but disperse at night, when fewer predators are about, to feed more solitarily.

Food storage, as discussed in Chapter 5, is a fall and winter practice both of migrants and of permanent residents that spend the winter in one place. Chickadees, White-breasted Nuthatches, and Blue Jays may begin storing food in October, but how much they later recover — before it has been stolen or they have forgotten where they put it — varies. Red-headed Woodpeckers are efficient acorn storers that recognize rivals in other species; they drive away all other woodpeckers, except the noncompeting Yellow-bellied Sapsuckers, and other acorn eaters like Blue Jays and Tufted Titmice.

Winter territoriality is common in both sedentary and migratory birds, but it is rarely as obvious as with breeding territories, since they are not usually announced with song or defended as vigorously. The Mockingbird is one of the few species that uses song to mark its territory; after the breeding season, young birds wander away or are driven off, the pair splits up, and each bird has a separate territory. The Red-headed Woodpecker also has an individual territory; the birds are most defensive of their acorn stores and sleeping holes, of which each bird has more than one.

Some species maintain group territories. Up to five White-breasted Nuthatches may jointly defend an area against outsiders; in Black-capped Chickadees the group may number a dozen individuals. Within these groups a hierarchical pecking order may develop, with some individuals always able to drive others away from a food source or favored roosting place. If the group includes a pair that bred on the territory earlier, these two usually outrank the fall arrivals. At the end of the winter, the breeding pair remains while the others disperse. Individuals often return to the same winter territory in succeeding years.

Dark-eyed Junco groups average a dozen birds and several members may roost in a single tree. During the day the group follows a fairly regular feeding route through its territory. Sometimes junco territories overlap and the two groups encounter one another, but there is little aggressive interaction; when each group moves on to its next feeding locality it is not joined by members of the other flock. White-throated Sparrows do not move in groups in winter, but at certain favored feeding sites like a bird feeder a similar rank order is maintained among all visitors. Males usually outrank females, but individuals with the little yellow patch in front of the eye always outrank those lacking it; even migrants visiting the site for the first time win encounters with winter residents that lack the yellow patch.

Groups of more than one species are a regular sight in the winter woods. The local chickadee species is generally the commonest bird in these groups; with the chickadees may be nuthatches, kinglets, a Brown Creeper, or one of the smaller woodpeckers. No two species compete for the same food, and group members do not repel newcomers of another species. The chief advantage of winter groups, whether mixed or single species, over solitary feeding is the easier detection of predators; another is that less energy is used in territory defense. Most of the group territory holders are smaller passerines, with the most enemies and least surplus energy to spend on aggression.

We see the same territorial strategies among long-range migrants and can make similar correlations between territorial behavior and diet. Some fruit eaters like the American Robin in Mexico roost in huge flocks and spread over the surrounding countryside each day to feed. Insect eaters, like the Northern Waterthrush, return as regularly to a winter site as to a breeding site, and defend it against other individuals. The roving mixed flocks of tropical fruit and nectar feeders are joined by small migrant insect eaters that have changed their diet. These tropical mixed flocks are usually made of 30 to 40 individuals of about ten species. Certain species, like the chickadees in temperate region flocks, are at the core of each group, and seem to attract other species; these nuclear species are usually gregarious among themselves and dully colored. For winterers that can make the dietary change there are several advantages to joining such flocks, and newly arrived birds have been seen carefully watching and following native birds to learn from them. Identification and localities of new food sources, plants to avoid, roosting locations, and predators are some of the things new arrivals learn more quickly when they join native birds.

Some tropical winterers never really settle down in one place.

Many birds wintering in Central America and northern South America move south and somewhat east during the first months after their arrival, stopping for as long as a few weeks to roam with a group of native birds, whose circular route keeps them within a limited area, and then moving on to join another group. In late December and January, many reverse their direction, and others later come north more directly via Honduras and Yucatan or the West Indies.

Many of the behavioral and dietary changes found in birds wintering in the American tropics are the result of such dense concentrations in a geographically small area. Most tropical winterers are in Central America and northern South America, a much smaller space than the North American regions from which the birds came. Only a few woodland birds like the Red-eyed Vireo and Blackpoll Warbler winter in the Amazonian forests of Brazil. In the Old World, the situation is quite different: many birds from Europe, Russia, and Asia winter throughout Africa south of the Sahara, a region very much larger than the part of the American tropics used by most North American birds. While American tropical winterers arrive at the end of a long rainy season when food is scarce, birds wintering in Africa arrive at the end of a short rainy season that has made insects and seeds abundant. One sign that competition is less severe in Africa is that winterers there do not join groups of local birds.

The concentrations of wintering birds in the American tropics create competition not only among individuals, but also among different species that do not meet when breeding. In Jamaica, where the Northern Waterthrush competes for ground insects with the Louisiana Waterthrush and the Ovenbird, it is confined to muddy lowlands; in Trinidad the Northern Waterthrush occurs alone, and extends its feeding places to rocky stream edges and dry forests.

Permanent residents are also affected by the winterers and transients. On Jamaica the two resident warblers are joined by 18 other species in winter. The local Yellow Warblers are restricted to the mangroves and woodlands close to the shore for this whole period; in April, after the winterers and migrants are gone, they expand to other forests to breed. In Colombia most wintering birds occur in the highlands, where the natural habitat is broken up by forest edges, second growth, and coffee plantations that provide lower vegetational levels than the humid lowland forests. These have less food and are more exposed to predators than the upper levels, but in winter many of the native birds feed extensively in them, returning to the higher levels after migrants have departed. The commonest residents here begin breeding shortly after the last migrants leave.

Finding a safe place to sleep at night is a particular challenge to birds in winter. In cold regions, the center of a dense evergreen or a tangle of vines near the ground is not significantly warmer than an exposed perch, but it is out of the wind and sheltered from precipitation, as well as being less conspicuous. Most small birds roost in such sites, but some larger ones like hawks and crows may spend the night high in a bare tree. In completely exposed habitats, birds like Rock Ptarmigans and prairie chickens each burrow a short tunnel into the snow, putting themselves out of sight and out of the wind; however, if sleet falls during the night and covers the surface, they may be sealed in the tunnel.

Hole-nesting birds often roost in holes as well, and winter hole users are usually permanent residents. In England, for example, resident Starlings roost in holes while winterers from Europe do not. Bluebirds that winter south of their breeding area roost in trees on all but the coldest nights. If a pair of woodpeckers winters on its breeding territory, the female may sleep in the nest hole while the male carves a new one for himself; young birds sleep outside, clinging to a tree trunk. As we have seen, each Red-headed Woodpecker zealously defends a territory containing more than one sleeping hole. The related Golden-naped Woodpecker of Central America and Colombia lives in a hole as a family unit; parents and young all roost in the nest hole until the following breeding season, when the young are driven out. White-breasted Nuthatches usually roost singly; occasionally two may share a hole, and as many as 29 have been found in a large tree cavity. Pygmy Nuthatches roost as a family, sometimes with other families; 100 were found in one hollow tree.

Some birds build special nests for winter roosting. Long-billed Marsh Wrens, permanent residents in the northwest, build individual winter nests of the same design as their breeding nests. Cactus Wrens also build individual nests, while in Central America various wrens make nests used by the pair or the family group. Several other tropical birds build winter dormitories, but the habit is not related to the degree of cold at night. Sedentary birds that stay in family groups and breed in covered-over nests are the most likely to winter in dormitories, but roving species and builders of cup-shaped nests never do. Dormitory and open roosters are both found in the coldest tropical mountain regions and in the warmest lowlands.

Since the winter movements of most birds are responses to changing food supplies, species with food sources that fluctuate are likely to be less regular in their winter habits. Even Black-capped Chickadees and Red-breasted Nuthatches, which are permanent residents or regular migrants in different portions of their ranges, sometimes move south in much greater numbers than usual. Occasionally a few Boreal Chickadees come south with the black-caps. In these years the black-caps and nuthatches are extremely common in the northern states during September and October, wintering farther south, but curiously never as abundant the following spring when returning north. Such periodic irruptions may be due to particularly successful breeding seasons that overpopulate the normal wintering area by late summer, or to food shortages. Sometimes one cause may follow the other, but it is hard to understand how chickadees can evaluate the fall and winter food supply by late August or September when, in irruption years, many have already begun moving south.

The irruptions of northern predatory birds like the Rough-legged Hawk, Snowy Owl, and Northern Shrike are better known. Each is tied to the population cycles of its winter prey, lemmings for the raptors and mice for the shrike. The rodent populations build up and suddenly drop every three to five years; in the same years the bird populations may also rise with increasingly favorable feeding conditions. When the prey population decreases, the birds must shift their diet or locality. During the autumn of the drop-off, large numbers move south in search of other prey, and by November they reach the northern states, where they find enough food to spend the winter. The Snowy Owl is more dependent on lemmings than the Rough-legged Hawk, so their invasion years do not always coincide.

More irregular invasions south of the breeding range are made by the finches like siskins, redpolls, crossbills, and Evening Grosbeaks that feed on tree seeds. Cones and alder and birch seed production fluctuate from year to year. When food is not available locally, the finches may wander hundreds or even thousands of miles during the fall and winter. Since each species specializes in a different food, few years produce invasions of all the "winter finches." Evening Grosbeaks and Pine Siskins regularly go farther south than the crossbills or redpolls; occasionally Evening Grosbeaks even reach Florida. The Pine Grosbeak is the least frequent or regular invader, and appears only in small numbers, not coming to feeders like most of the others.

True hibernation, such a common winter strategy of amphibians, reptiles, and mammals, is known for only one bird species. In 1946 a Poor-will was found sleeping in a rock crevice in the Chuckawalla Mountains of California. Its body temperature was only 60° F and its breathing rate was hardly measurable. The bird was banded, and was found hibernating in the same crevice the following three winters. Hibernation has not been found in any other caprimulgids; most live in or migrate to warm climates.

Behavior akin to hibernation has been found in Finnish Ring-necked Pheasants. In midwinter some stay in their roost tree for 40 to 42 days, totally inactive and insensitive to disturbances; since these roosting periods are not restricted to cold spells, it is believed that they are regulated by some internal factor.

There are several interesting ways to study movements and behavior of birds in winter. The Christmas Counts sponsored by the National Audubon Society are an invaluable source of data on the population changes of each species in an area, centers or shifts of population, frequency of invasions, and occurrence of stragglers out of their usual winter range. Each count surveys the same area, bounded by a circle fifteen miles wide, year after year. The more observers on a count, the more thoroughly the area is covered, and the more accurate the data become. You can perform a real service by participating in a count; if no count takes place in your area, the National Audubon Society can tell you how to start one.

Bird counts of one area made on a regular basis through the winter provide other valuable kinds of information on how each species is affected by changes in weather, dietary shifts, arrivals and departures of nonresidents, formation or breakup of territories, and other aspects of behavior. If you follow the same route on each survey and in succeeding years, you will be able to draw more valid conclusions than from irregular visits to different sites.

Feeder watching is entertaining and informative. Chapter 14 gives details on different types of feeders and appropriate food. At a feeder you can see how each species handles food, which ones go off to store it, which species and individuals give way to others, and how often each comes to feed. Correlations of feeder activity with day length, brightness of daylight, and temperature are easy to document and fascinating to analyze. Studies like these will give you insights into how birds cope with the challenges of winter.

Chapter Twelve

Cattle Egrets are now as at home with American dairy cattle as with African water buffalos.

Distribution

THE DISTRIBUTION of birds around the world has long been studied by both ornithologists and ecologists, because so much can be learned from an understanding of why birds occur where they do: the ornithologist uses distributional studies to help understand a bird's life history, its relationship to other members of the family, its evolutionary history, and the family's place of origin; the ecologist studying distribution is using birds as demonstrations of general ecological principles of life zones, plant and animal communities, colonization and dispersal patterns, effects of climate, and other factors affecting all forms of life.

Why some species or families are widespread while others are restricted to a small area is the first problem the student of distribution must consider. Some of the families having a nearly worldwide distribution include the ducks and geese, hawks, rails, owls, gulls, pigeons, swifts, swallows, and thrushes. Certain species live almost all over the world: the osprey is found throughout the northern Temperate Zone, into a few tropical areas, Australia, and New Guinea; the Barn Owl inhabits Australia, the East Indies, Africa, all but the northernmost areas of Eurasia and the Western Hemisphere, and many oceanic islands; and the Common Gallinule, not the type of bird you would think of as a strong flier or great traveler, occurs in most of the Western Hemisphere, Africa, Eurasia, and some oceanic islands, while other families are confined to only one continent, a series of islands, or even a single island. Within North America, for example, two hummingbirds — Anna's and Allen's — have narrow breeding ranges on the west coast, and the Kirtland's Warbler nests only in jack-pine country in a few counties of lower Michigan.

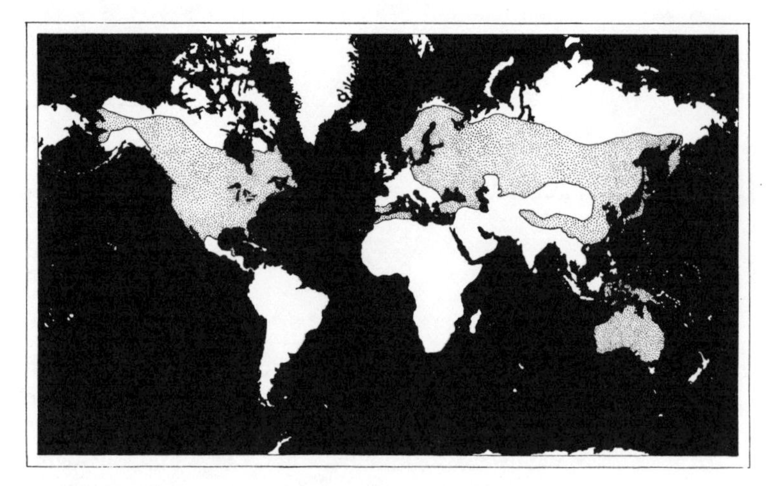

The breeding range of the Osprey. In winter it is also widespread — occurring from the southern part of the breeding range south through much of Central and South America, Africa, and Asia.

Factors Affecting Distribution

Clearly, birds that have evolved very specialized habits and exploit narrow environmental niches are less capable of dispersing over wide areas from their center of origin than are more adaptable or more mobile forms. Precisely what enables or prevents expansion of range varies from group to group, but some basic patterns are easy to see. Birds can fly, and therefore pass over what would be barriers to the dispersal of animals that travel on the surface of the earth — in migration many birds fly over large areas of unsuitable habitat; however, migration is a two-way movement between localities where the bird is an established nester or winterer, and does not involve any spread in the species' usual range. In cases of true dispersal, geographic features often limit spread — large stretches of water are barriers to land birds, seabirds may avoid going over land bodies, mountain birds don't cross plains, and so forth. It is important to remember that some present barriers did not always exist: during the Tertiary Period, from one million to sixty million years ago, when birds were abundant and many modern species already existed, there was land between Alaska and Siberia, over which many birds crossed in each direction; elsewhere, mountains that now split a bird's range may not have existed. Continental drift may have played a role in the present distribution of some particularly old families and orders, but the continents are believed to have begun drifting apart before the evolution of most modern groups of birds.

Some barriers are more effective than others — an exhausted land bird blown over the sea may perish if it drops to the water before encountering land, but a mountain species can rest and probably

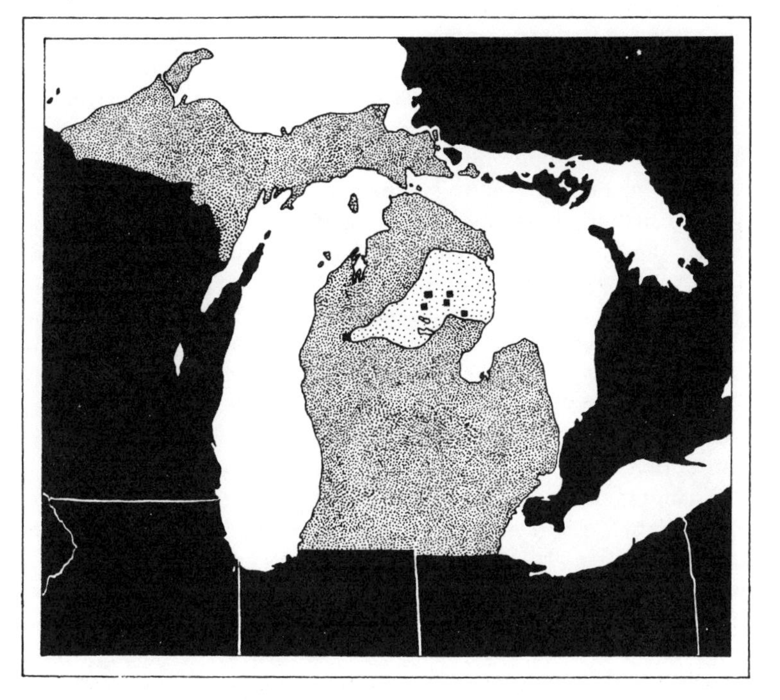

The breeding range of the Kirtland's Warbler. It winters in the Bahamas, and is rarely seen on migration between the two regions.

breeding range

recent breeding sites

feed if it lands in a plain before reaching other mountains. In other cases, the barrier of land or water is more psychological than real: at the Isthmus of Panama, seabirds like Brown Pelicans and frigatebirds regularly fly across the fifty miles of land, while the Brown Booby, an equally powerful flier, never does, and distinct subspecies are found on each side of the isthmus; in the Amazon region, rivers separate races of certain strong-flying species; and in North America non-migratory birds like the Ruffed Grouse and Screech Owl are absent from large islands of suitable habitat separated by as little as a mile from areas where they are present.

Other barriers reflect more specific environmental requirements. Changes in ocean temperature that affect the food species have prevented certain seabird expansions: the belt of warm water across the mid-Atlantic has prevented albatrosses from expanding into the North Atlantic, while cold currents coming from the north and south on the American shores of the Pacific have prevented the Royal Tern from nesting beyond California and Mexico. The Brown Pelican does not occur on the Atlantic coast south of the Amazon River because it has never crossed the broad mouth of the Amazon, where the water is too muddy for a pelican to see its prey — once past the Amazon, the habitat would again be perfectly suitable.

The distribution of some species parallels a particular food source exactly: the Everglade Kite and the Limpkin both feed exclusively on snails of the genus *Pomacea,* and are uncommon in Florida, at the very northern edge of the snail's range, but are found throughout the American tropics south to Argentina, wherever the snails occur. Similarly, the Common Jay of Eurasia feeds mainly on acorns, and its range closely parallels that of oaks.

Competition with other species for a particular niche also affects distribution, especially at the southern end of Northern Hemisphere bird ranges, because, approaching the tropics, there is generally a large number of species each exploiting a more finely divided environmental niche. The Belted Kingfisher, for example, encounters in the southern part of its winter range two other kingfishers, one slightly larger and one slightly smaller, preying on the same size range of fishes the Belted Kingfisher takes; there is enough food available in winter, but not for all three species to breed, so the Belted Kingfisher nests only north of this area.

Modern Changes in Bird Distribution

Some modern cases of changes in bird distribution illustrate other factors affecting distribution. As the following examples will show, essential to any increase in a bird's range is a vacant niche that it can exploit; sometimes such a niche has always been available, as the coast south of the Amazon has been for the Brown Pelican, which has never crossed the barrier of muddy water between suitable areas, but at other times, environmental changes make a new area usable.

Man has been one of the sources of rapid environmental change, and the distribution of some birds reflects this. In North America, man's habitat alteration has reduced the areas usable to many birds, but (except for those species he has driven to extinction) their distribution is unchanged; other species have increased their range because of these changes: the Brown-headed Cowbird was originally called the "buffalo bird" because it lived with the great herds of bison on the prairies of the Midwest, but with the clearing of vast areas in the East and the presence of cattle there, the cowbird has spread to the Atlantic coast.

Man is directly responsible for changes in distribution of the Northern Fulmar and certain gulls. The fulmar, a northern ocean shearwater, has profited by commercial fishing methods, which gut and clean fish at sea, throwing the refuse overboard; this new food source has created a population expansion, forcing fulmars to colonize new areas — while in 1877 there was one colony in the British Isles, colonies now exist all along the British coast and in northern France. Curiously, in the western Atlantic, fulmar colonies have not spread south of northern Labrador. The Herring Gull, which in 1920 nested no farther south in North America than central Maine, now nests as far south as South Carolina, and the Great Black-backed Gull, originally also a northerly species, has spread almost as far south; both have expanded their ranges because modern food sources such as garbage dumps, sewage, and dumped fish wastes have allowed more to survive through the winter months, when food was earlier in shortest supply. With more gulls surviving, existing colonies are filled, and some gulls are forced to expand to new areas.

Direct introduction by man has of course changed the distribution of many birds. Almost everywhere English colonists have settled, they have brought birds from home and released them. Most attempts at introduction (like the 1890s American movement to introduce all the birds mentioned in Shakespeare) fail, but the few "successful" introductions have sometimes had harmful consequences for the native birdlife. The best known North American introductions are the House Sparrow and the Starling. House Sparrows, not true sparrows, but members of the Old World weaver finch family, were first released in Brooklyn, New York, in 1850. They spread rapidly, especially wherever man provided abundant food around farms and stables, and are now found throughout North America and into Central America as far south as El Salvador. The Starling was first successfully introduced in New York City in 1890, and it too has spread all over North America and into Mexico. Both are successful because they have little or no competition for food, and are more aggressive than native birds with which they often compete for natural or manmade nest sites; bluebirds, Tree Swallows, Purple Martins, Great Crested Flycatchers, and woodpeckers are all victims of one or both species.

Several game birds have been introduced in North America; the most widespread is the Ring-necked Pheasant, but the Chukar and Gray Partridge are also locally common. The Mute Swan on the eastern seaboard, European Tree Sparrow around St. Louis (a close relative of the House Sparrow with apparently too specialized habits to spread far in a new environment), and Skylark and Crested Myna in British Columbia are some of the other introduced species of long standing. Recent introductions of exotic parrots, bulbuls, and tanagers in south Florida may prove harmful to the native birdlife, as introductions so often are. In Hawaii many of the specialized native birds have become extremely rare or extinct, in part from competition with introduced species from Asia, North America, and Europe.

An unusual introduction is that of the House Finch, of western North America, in the northeast: individuals of the California subspecies were first noticed on Long Island in 1941; they were probably released by Brooklyn pet dealers informed that the species could not legally be sold as a pet. House Finches nested first on Long Island and then spread into the New York City region; in New York City itself they thrive wherever there are planted terraces or backyards — their song can be heard over traffic on many busy avenues. The species has since spread along the east coast from Massachusetts to the Carolinas. Unlike most deliberately introduced species, it is a native of North America and has not been found to compete even with its close relative, the Purple Finch.

Other recently documented range expansions may be due to climatic changes: the Tufted Titmouse, Mockingbird, and Cardinal, all rare north or east of New Jersey before the 1950s, are now widespread permanent residents into northern New York and New England; a succession of mild winters and an increase in feeding stations are considered responsible for these expansions.

Some range expansions due to population increases are difficult to explain. Until the early 1940s, for example, the Glossy Ibis bred within North America only in Florida, then its range expanded very rapidly, so that by 1961 the Glossy Ibis nested on Long Island, where it is now common in at least six colonies. In 1973 Glossy Ibises began nesting in Maine; why the ibis expanded its range so suddenly and rapidly is unknown — had the expansion taken place before observers were present to record it, ornithologists would have assumed the ibis' wide distribution to be of long standing. Similarly, other species whose distribution we consider stable may in fact have only recently established their present range.

Other changes in bird distribution are the accidental results of weather. In 1937 a storm blew a flock of migrating Scandinavian Fieldfares (a species of thrush) west to the coast of Greenland, beyond their normal range; the birds found conditions acceptable on the southern coast of Greenland and have been permanent residents there ever since.

The most spectacular case of range expansion observed by man is that of the Cattle Egret. Originally a native of Europe, Africa, and Asia, the Cattle Egret first appeared in northern South America during the 1930s, having presumably been blown over from Europe or Africa. Feeding on insects in fields, where it often follows cattle that flush up its food, it found no competition for this niche in the New World, and spread rapidly through the West Indies to North America, reaching Florida in the early 1940s. It began nesting in Ontario in the mid-1960s, reached California by 1964, and has occurred as far north as Newfoundland.

The World's Zoological Regions

Besides the very rapid changes in bird distribution observed and described over tens or hundreds of years, ornithologists try to ex-

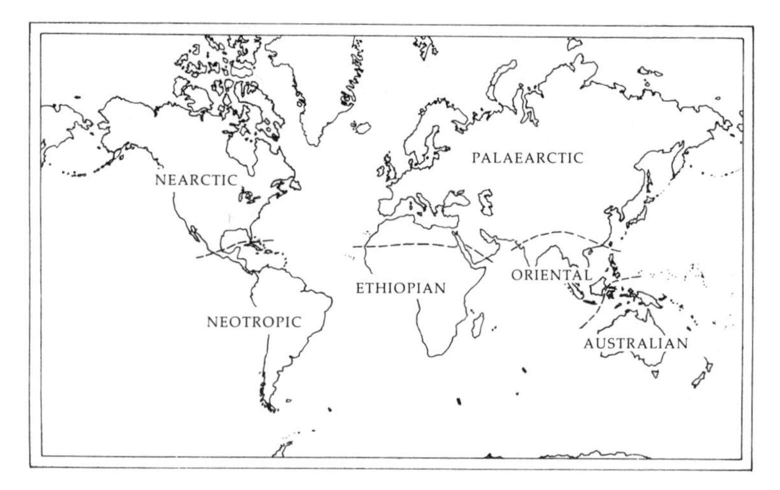

The major zoological regions of the world. The Holarctic is divided into the Nearctic and the Palaearctic, between Greenland and Iceland on one side and between Alaska and Siberia on the other.

plain the long-term changes and the larger, more stable patterns of distribution. By examining the distribution of entire families and orders of birds, and correlating it with features of geography, climate, vegetation, and other basic environmental indicators, scientists have divided the world into five major zoological regions that apply to many other animal groups as well. There is, of course, overlap between each of the regions, and some areas within the large regions are so distinct that they have been considered subregions. Each habitat has its own set of birds, but within a region the same community of birds is apt to be found wherever that habitat occurs.

1. *Holarctic (Palaearctic and Nearctic) Region.* This includes nearly all of the Northern Hemisphere, divided into two subregions, the Palaearctic (Eurasia and North Africa) and the Nearctic (North America and part of Central America). These were once connected by land over what is now the Bering Sea, between Alaska and Siberia, and many small birds that might not now cross the Bering Strait could then expand their range overland into the other continent. In the northern parts of the Holarctic, habitats are nearly the same throughout the region, with polar seas, tundra, and, farther south, conifer forests extending in parallel belts across Canada, Siberia, and north-

ern Europe, and the same, or very similar, bird species occurring within each belt in both subregions. Many of the same loons, alcids, and waterfowl are found around all of the polar seas, and the tundras share many species of shorebirds, hawks, owls, and some passerines; a few of these are the Black-bellied Plover, Gyrfalcon, Snowy Owl, and Snow Bunting. Other tundra species, such as the small sandpipers called "peep" in North America and stints in Europe are very closely related and at some seasons indistinguishable.

South of the tundra, in the boreal forests of conifer and birch, there is a greater mixture of birds found through much of the Northern Hemisphere, like the Goshawk, Great Gray Owl, Brown Creeper, kinglets, redpolls, and crossbills, with families restricted to one of the subregions. Most of the restricted families, like the flycatchers and wood warblers in the North American conifer forests, are highly migratory, leaving for the tropics each winter, while the Holarctic species travel less, if at all.

In the deciduous forests, grasslands, and deserts farther south in the Holarctic region, the number of species found in both subregions decreases. While the northern habitats occur in continuous belts around the hemisphere, making expansion from one subregion to the other easy, Nearctic deserts and grasslands, for example, are widely separated from those in the Palaearctic by stretches of habitat inhospitable to specialized desert or grassland birds.

Passerines especially are usually different species and often from different families within the subregions of the Holarctic, because to them ocean, tundra, or boreal forests are more of a barrier than to large birds like waterfowl, hawks, and owls. Moreover evolution of passerines began more recently and its rate is faster than that of nonpasserines, so that birds in similar but widely spaced habitats would evolve separately; many of the specialized habitats themselves, such as the western deserts and grasslands, are of fairly recent origin and the birds found in them are recently evolved specialists from families primarily found in other habitats, and often from families that have recently expanded from zoological regions farther south.

2. *Neotropical region.* The Caribbean islands and Central and South America form the most easily separated ornithological region because twice in the past, from 50 to 70 million years ago, and 2 million years ago, most of Central America was under an ocean, and South America developed several families that have never spread north. Some, like the toucans, are well known; other small families

are still little known even to ornithologists. Neotropical habitats include lush forests receiving heavy rainfall, the high Andes mountains, broad marshes and grasslands, mountain and coastal deserts, and a temperate region.

Because plant life is so abundant and varied, the Neotropics are characterized by an enormous variety of bird species, each occupying a very specialized niche. Colombia alone, the size of Texas and Oklahoma, has 1532 bird species, more than twice the number found in North America north of Mexico. Instead of the approximately 50 species found in boreal forests, Neotropical forests have hundreds of species living in six levels from the forest floor to the tree canopy a few hundred feet above. Few of these species, however, are as widespread or common as many Holarctic birds. Rather than the few abundant species exploiting a fairly uniform environment, as in northern parts of the Holarctic, there are many less common species, each exploiting a small niche in more varied and easily partitioned habitats.

In addition to the families restricted to the Neotropics, there are of course many families shared with North America, such as the mockingbirds, wood warblers, and blackbirds, and others of worldwide distribution, especially water birds, hawks, owls, woodpeckers, and some passerine families, including the swallows and thrushes. However, it is interesting to note that while there are several Old World families in South America, no family of South American origin has spread to the Old World, perhaps indicating that temperate families can adapt to the tropics more easily than tropical families can to temperate regions.

A few Neotropical families, like the parrots, barbets, and trogons, are found elsewhere primarily in other tropical regions. Historical and fossil evidence shows that these families once occurred north of the present tropics — the Carolina Parakeet, exterminated early in the twentieth century, once lived in Virginia and the Ohio Valley; fossil parrot remains have been found in Nebraska, England, and Germany, and a few parrot and trogon species still occur in northern Mexico and Arizona. Perhaps these families were once found throughout the Northern Hemisphere, until dry and cold periods reduced the tropics and their associated birds to their present discontinuous range.

3. *Ethiopian region.* The Sahara desert and those of Syria and Arabia separate Africa and the Arabian peninsula from the southern

portion of the Palaearctic. Besides the largest deserts in the world, the Ethiopian region includes grasslands, some interspersed with trees, rain forests, and wetlands. As in the Neotropics, birdlife is abundant and varied, from the Ostrich, found in grassland and arid (but not desert) regions, to tiny, iridescent sunbirds, the ecological replacement of the New World hummingbirds.

Unlike those in the temperate region, the grasslands of Africa have a tremendous variety of bird life, each species occupying a more specialized niche, in the typical tropical pattern; the birds of the East African grassland, which is dotted with trees and scrub, are remarkably easy to observe, and the national parks of Kenya, Uganda, and Tanzania have become major tourist attractions for their birds as well as for the large mammals.

Five small families occur only on Madagascar, which has been separated from Africa for millions of years, and a number of mainland families have never reached the island.

4. *Oriental region.* Unlike the Neotropical and Ethiopian regions, separated from most others by oceans or deserts, the Oriental region has fewer geographical barriers and shares many bird families with adjacent areas. The Himalayas are an effective barrier to tropical birds in India, but east and west of the mountains, lines are less distinct; the Philippines and the islands of Indonesia, which were once connected to southeast Asia, are part of the region, while New Guinea and Australia, which have been separated from Asia much longer, form another region.

The oriental region includes tropical lowland forests and others on steep mountain slopes, coastal mangrove swamps, and flat, dry areas in the central plain of India. In addition to its varied resident birdlife, the Oriental region is the winter home for vast numbers of Palaearctic migrants, some of which fly over the Himalayas, as well as some shorebirds that breed in Alaska.

5. *Australian region.* Australia, New Guinea, and New Zealand have been islands separated from the nearby Oriental region for so long that they have evolved a birdlife distinct from the other regions and from each other. Confined to Australia are the emus and lyrebirds; New Guinea has the birds of paradise, and New Zealand has the kiwis. Some families, including the parrots, pigeons, and kingfishers, of wide tropical or world distribution, are most numerous and diverse in the Australian region, while other widespread families, such as the finches, do not occur there at all.

In the lush mountain forests of New Guinea, the variety and specialization approach those of Neotropical forests. Some species restricted to certain altitudes, where they are isolated by valleys or peaks from other populations of their kind, have been frequently studied as examples of patterns in local distribution and in the evolution of races and species.

Seventy percent of Australia's land area is desert, where rain is irregular, but sudden and briefly abundant. Many birds have evolved unusual adaptations for survival here; they wander hundreds of miles in search of water and, unlike birds of all other deserts, whose nesting is restricted to a certain season, begin breeding almost instantly at any time of year if it rains — at least one species, the Black-faced Wood Swallow, is known to begin gathering nest material at the mere sight of a rain cloud.

New Zealand, small, cool, and removed from other land areas, has a limited but highly distinctive native bird life. Compared with

A moa, 10 feet tall, member of a family confined to New Zealand that became extinct several hundred years ago.

other regions it has very few passerines, which presumably have had difficulty reaching or settling here, although several niches remain to be exploited. The British introduced about 25 bird species here during the nineteenth and twentieth centuries; some have caused the decline of native species by competing with them, while others have had no harmful effect, filling vacant niches. New Zealand's most unusual birds were the moas, heavy, ostrich-like flightless birds ten feet tall; they became extinct several hundred years ago, after the arrival of man.

Where Different Bird Families Originated

With a knowledge of the present distribution of birds within any region or throughout the major regions, it is possible to theorize on the origin, evolution, and previous distribution of families and even species. When fossil evidence is scanty or lacking, present distribution may be the only source of information; in these cases, however, one must be especially careful, since judgments based only on present distribution can be misleading. From fossil remains we know that the New World vulture family was once found in the Old World, and the Old World vultures, part of the hawk family, were found in the New World, but whether each family is now confined to its original home or survives only in what was once a range extension is not known.

The first requirement of bird distribution studies is an accurate knowledge of systematics — without this, one can make serious mistakes. For example, in 1924 the wren family was considered to have 48 species and subspecies in the Old World, as well as the 59 species in the Western Hemisphere, but just over twenty years later, ideas of bird relationships had so changed that all but one of the 48 Old World forms were considered part of another family, the babblers. Similarly, the Wrentit of California, Oregon, and Lower California, formerly considered a separate family of one species restricted to North America, is now recognized as the only New World representative of the babbler family. Obviously, changing ideas of family relationships caused major revisions in our ideas of how the wrens and babblers got where they are today.

Families now so widely dispersed that it is impossible to say where they originated include the oceanic birds, shorebirds, fresh-water birds, hawks, swifts, woodpeckers, and swallows, but the small families, especially those of the relatively rapidly evolving passerines, that are now found only in one region or one island like Madagascar or New Zealand are assumed to be at their center of origin. Sometimes judgments can be made about the families that now inhabit more than one region, because it is generally believed that the region where a family is present in the greatest variety and with the most unusual types is its place of origin, since the family has had more time to evolve there, while regions with only a few members of the family may be more recent expansions.

Applying these principles to the birds of North America, we see that our birdlife is a mixture from many regions. The loons, for example, are strictly Holarctic, but we cannot say whether originally Palaearctic or Nearctic. For the alcids, also strictly Holarctic, it is more appropriate to speak of the ocean rather than the continent of origin; they are most diverse in the North Pacific (both Palaearctic and Nearctic) and probably spread from there to the Atlantic. Other families found only in North America in the New World, like the titmice, nuthatches, and creepers, occur in greater variety in Old World regions. To these families and those that also reach the Neotropics, but are still most varied in Old World regions, such as pigeons, cuckoos, kingfishers, larks, jays, and pipits, we attribute Old World origin; the pigeons and kingfishers are most numerous and diverse in the Australian region, and may have originated there, but reached the New World from the Palaearctic.

No family is now exclusively Nearctic, but several definitely originated in that area, and most of those have since spread into the Neotropics, a few into the Palaearctic. The wren family, now widespread throughout the New World, has one species that has spread across the entire Palaearctic — the Winter Wren. The grouse family has not spread south to the Neotropics, but now occurs throughout the Holarctic, with many exclusively Palaearctic species — that new species have evolved after grouse reached the Palaearctic suggests that grouse have been present there longer than the wrens, whose one representative is only subspecifically different from the Nearctic form.

Among the familiar New World families that originated in North America are the turkeys, mockingbirds, vireos, and wood warblers. The turkey family has two species: the Common Turkey — ancestor of the domestic bird — which once occurred wild throughout the deciduous wooded portions of North America, and the Ocellated Turkey, restricted to the Mexican region, the southernmost part of the Nearctic. Similarly, mockingbirds and vireos are most diverse in Central America and may have spread both north and south from there. Although varied in the Neotropics, wood warblers are believed to have originated in North America, because of the larger number of genera and species that breed only there.

Families of clear Neotropical origin in our birdlife include the hummingbirds, flycatchers, blackbirds, and tanagers, all very large families, with (except for the blackbirds) fewer than 10 percent of their species found north of Mexico. The hummingbird family, for example, includes 320 species, of which one is found in eastern North America, twenty in western North America, fifty in Mexico, and one hundred in all of Central America. The flycatchers, blackbirds, and tanagers similarly become more numerous as one approaches the Neotropics.

Two large passerine groups remain to be discussed, the thrush family and the finches, which, as we saw in Chapter 3, may actually be two families. The thrushes are of Old World origin, but none of the New World species are also native to the Old World (except the Gray-cheeked Thrush, which has extended its breeding range into Siberia, and the Wheatear, which has expanded from Siberia to western Alaska; each returns to its continent of origin in winter). Some entire genera are of New World origin, indicating a long period of evolution in North America. The finches are now worldwide, except in the Australian region. Only the original homes of certain groups within the family can be guessed; the typical brown-streaked sparrows probably originated in North America, the goldfinch-redpoll group in the Palaearctic, and the cardinals in the Neotropics.

Biomes of North America

Within any of the zoological regions of the world it is possible and often useful to divide the area into smaller units based on underlying environmental factors like climate, soil type, and vegetation. These regions are called biomes, or biotic communities. In North America there are five major biomes — tundra, coniferous forest, deciduous forest, grassland, and desert. Within each biome, of course, are smaller ecological units — the deciduous forest biome, covering hundreds of square miles of eastern North America, includes meadows, marshlands, and second growth, as well as mature forest. Each type of growth is a separate ecosystem, with some birds found in no other habitat and some flexible species that can utilize several.

1. *Tundra.* The tundra is the northernmost land in North America, with ground permanently frozen, except for the surface, which thaws out during the short summer. Trees, except willows only a few inches tall, cannot grow here, but lichens, grasses, sedges, and wildflowers are abundant. The annual thaw creates many ponds and puddles, breeding grounds for vast numbers of insects on which many birds feed. In June and July there is continual light, enabling birds to go through the nesting cycle more quickly in the short time available before the cold weather returns. During winter there is little light, but also little snow; some birds — ptarmigans, Gyrfalcon, Snowy Owl, and Raven — move south only if food becomes scarce; others, like the waterfowl, shorebirds, and the Snow Bunting leave the area entirely. Some of the shorebirds, as we have seen, go as far as the pampas of South America where they find a similar grass and pond environment.

2. *Coniferous Forest.* In the southern portion of the tundra are stunted spruces and balsam firs. Continuing south, the trees grow taller and crowd out the tundra, forming a broad belt of boreal forest that runs all the way around the world; some of the birds restricted to the North American portion are the Spruce Grouse, both three-toed woodpeckers, and the Gray Jay. In summer the abundant insect life makes these forests attractive to several flycatchers and warblers, as well as to the finches that specialize in cones and birch catkins but feed insects to their young; many of these species breed quite far south in the Rocky Mountains, where the conifer forest occurs at high altitudes.

3. *Deciduous Forest.* Deciduous woodlands, the dominant forest type in the eastern United States, have the greatest variety of birds of any North American biome. The forest is no longer unbroken, as is the coniferous forest, nor is it uniform. In the north it is made up of maple, beech, and hemlock (a conifer); the central portion is primarily oak and hickory; and the southern coastal states have forests of pine species not found farther north. Each type of forest has some birds restricted to it, while other birds inhabit forests of the entire biome; many birds, of course, live only in the nonforested areas of

the biome, but a meadow species, for example, is still considered part of the deciduous forest biome rather than the grassland.

4. *Grassland.* This includes only the area that was always plain, not cutover eastern forest that is temporarily grassland, and covers the central portion of North America from the Rio Grande to the Canadian prairie provinces. Recent plantings of trees around homes and settlements have allowed several deciduous forest species to spread west, but the grasslands themselves have a low density and small variety of breeding birds, including the Burrowing Owl, prairie chickens, Western Meadowlark, Dickcissel, and several sparrows.

5. *Deserts.* Deserts have little rainfall and poor soil that cannot hold the rain when it occurs. Plant life has adapted to absorb water quickly and retain it, or survive with very little; since vegetation is never dense, it supports animal life only at low concentrations. In addition to lack of water and sparse plant life, desert birds must contend with wind and extremes in hot and cold temperatures every day and night. The desert specialists include the Elf Owl; the Gila Woodpecker, which climbs up cactus the way other woodpeckers climb trees; the Roadrunner; the Cactus Wren; and a few thrashers and sparrows. Some wide ranging birds, not restricted to any biome, are also found here, including several predators and scavengers — the Turkey Vulture, Red-tailed Hawk, Golden Eagle, American Kestrel, and Raven — that seem able to find food anywhere.

Each biome, of course, can be broken down into smaller units of habitat based on vegetation, water, soil, altitude, average temperature, etc. It might be interesting to examine each distinct habitat near you — forests with different kinds of trees, parks, farms, marshes, young trees and shrubs — for the source of the birds found there. Within each habitat how many birds are of Old World, Nearctic, Neotropical, or unknown origin? Combining all the habitats in your area, how many birds come from each region? Whether species are at the edge of their range, and if their ranges seem to be growing or shrinking, are important questions that always need attention and depend on regular observations made all over the country.

Chapter Thirteen

The Dodo of Mauritius Island, about 2½ feet tall and weighing up to 50 pounds, was an easy victim to sailors and animals released on the island. It disappeared soon after 1680.

Conservation

THE GROWING awareness that we share the same environment with all plants and animals and that what affects any living creature affects many others, including man, has led to a new interest in wildlife conservation. Extinction is, of course, part of the natural process of evolution, and millions of species have disappeared from the earth since life began, but this is no justification for the highly unnatural ways in which human activities have eliminated entire species. Most recorded bird extinctions have been due to a fairly simple combination of direct persecution by man — and his introduced agents like rats and pigs — and habitat destruction. These factors are still operating today, together with new causes. Conserving populations now very much reduced is a complicated affair, and too often we only become aware of the problem or devise a possible solution when it is already too late.

There are many reasons for conserving wildlife, from the critically endangered species to those so widespread that we could not conceive of their ever becoming rare. Each is the result of a long evolutionary process that can never be re-created; the Whooping Crane and the California Condor are the descendants of forms originating some fifty million years ago, and, as anyone who has seen either of them would agree, the wonder and diversity of the earth would be reduced without them. Less glamorous species are still equally deserving of protection. Birds also have economic, sport, and health values, as described in Chapter 2. With proper management, they can continue making their direct contributions to human welfare. Finally, birds serve as indicators of environmental problems that affect humans along with other forms of life. The manmade chemicals that have reduced or destroyed the reproductive abilities of certain birds are also in our bodies, where their effects are still unknown.

We are beginning to see some of the dire warnings made by environmentalists about exhaustive agricultural processes, mis-sited dams, clearcutting of forests, badly laid pipelines, and wasteful energy practices come true. Habitat altered so that it is unusable by birds is not necessarily put to a use more productive to man. As Jean Delacour wrote when president of the International Committee for Bird Preservation (now the International Council for Bird Preservation), "We play on earth a leading role, which should not be that of a villain."

Conservation is now a global problem. Every region has its endangered native wildlife, and in additon, we all share responsibility for the welfare of life in other parts of the world — pesticide residues have been found in people tested far from places where they have been used, and birds must be similarly affected. There is little hope for the restoration of the Peregrine in eastern North America as long as it can pick up massive quantities of the pesticides still used in South America, where it winters. No species can survive long even when fully protected in one country if it is over-hunted or its habitat destroyed in a nation it visits at another season. Nor can a bird last long if it is hunted to supply a demand for feathers or pets in another part of the world. A partial response to these problems has been the creation of covenants between the United States and some of the nations with which it shares birds; there are now migratory bird treaties with Canada, Mexico, and Japan. A treaty with the Soviet Union is currently being drafted. These treaties protect or regulate the take on birds that pass through the two countries, agree that protection is desirable for endangered species, and call for the exchange of research information. There is also a series of United States tariff and endangered species acts that ban the importation of certain wild birds or their products. The Convention on International Trade in Endangered Species of Wild Fauna and Flora, effective July 1, 1975, prohibits or restricts the commercial trade of rare and endangered birds among all ratifying nations, of which the United States is one.

The history of extinct birds has the fascination of all stories to which we respond, "If only they had realized what they were doing!" The story will only be instructive, however, if we realize what we are doing today. A few regular themes run through the accounts of extinct and very rare birds. Nearly all extinctions have occurred where European man arrived in an area that had been undisturbed long enough to allow the evolution of birds with very specialized ecologies. Many had lived so long on predator-free islands that they had lost both the need and the ability to fly. These birds were especially easy victims of direct persecution by man and the animals he brought with him, or to habitat destruction. Greenway's definitive *Extinct and Vanishing Birds of the World* (revised 1967 edition) lists 151 forms of birds known or believed extinct since 1680. About 90 percent of the extinct birds lived on islands; most of the others were from North America.

The first bird known to become extinct in recorded history is the Dodo of Mauritius Island in the Indian Ocean. It disappeared soon after 1680. Flightless, with a nine-inch-long, hooked beak, and weighing up to 50 pounds, Dodos probably lived on the ground in forests. Although the Dodo was a case of extreme evolution, the story of its extinction closely parallels many later ones. Mauritius was discovered by the Portuguese about 1507; in the sixteenth and seventeenth centuries Dutch, French, and British ships regularly stopped there, taking large supplies of Dodos for provisions and leaving behind pigs, monkeys, cats, and rats that easily killed the birds or destroyed their eggs. The long list of extinct and very rare ground-nesting island birds, including shearwaters, cormorants, ducks, quail, and rails, owe their status to similar patterns of hunting or, more frequently, persecution by introduced animals. Pigeons and parrots as well were used for food by visitors to islands all over the world; parrots were also frequently taken as pets. Cats, rats, and the mongoose may be entirely responsible for the extinction of so many ground-nesting and low-nesting passerines, while human destruction of forests or introduction of goats that ate all the vegetation have caused the decline of other island birds.

Extinctions in North America are slightly different, in that some island birds may not have been common even before the arrival of European man and were restricted to localities with few places of refuge, or were flightless or poor fliers. Some of the now extinct or very rare North American birds were both abundant and widespread, but it may still be said that they all had had particularly specialized habitat requirements or behavioral traits: the Passenger Pigeon rarely bred successfully except when in very large colonies, and both the Carolina Parakeet and the Eskimo Curlew (a bird not quite extinct) had the habit of returning to the place where their companions had just been shot.

The Great Auk, a large flightless alcid, illustrated on page 31, was the first North American species to become extinct. Perhaps due to hunting, its range gradually shrank over the last few thousand years — auk bones have been found in Scandinavian settlements dating from 6000 B.C. to 1000 A.D. — to a few islands in the North Atlantic. Funk Island, off Newfoundland, was the last large colony. Sailors regularly rounded the birds up or drove them onto their boats to slaughter them for food, fish bait, oil, and the feathers, which were used for bedding. Smaller numbers were taken by settlers on the coasts visited by auks in winter. (They occurred regularly south to Cape Cod.) Sometime in the early nineteenth century the Funk Island colony was wiped out; the last Great Auks were two captured on an islet off Iceland in 1844.

The Passenger Pigeon was once probably the most abundant bird in the world — in the nineteenth century there were single flocks estimated at over a billion birds, more than the entire population of any species known today. They usually lived and traveled in these huge flocks, breeding in the hardwood forests that occurred from the Atlantic coast to the Great Plains and feeding primarily on nuts and berries. Nesting colonies a hundred miles in length were recorded in the unbroken forests, although others nested in groups down to a few pairs where habitat was more scattered. The large flocks were subject to continual slaughter on the breeding grounds (in 1861, 14,850,000 were shipped from one Wisconsin colony), on migration when they "darkened the skies," and wherever they wandered in winter looking for nut and mast crops. Persecution at the colonies may have caused desertion of eggs or young, further reducing the population. In New England, nesting colonies of up to 5 acres existed until 1851, but the surviving population was increasingly centered in the western Great Lakes states, where their ultimate decline was rapid. The last large colonial nesting took place in Wisconsin in 1885, no nests at all were found after 1894, and the last wild Passenger Pigeon was shot there in 1899. The last surviving individual died in the Cincinnati Zoological Garden on September 1, 1914. Slaughter alone cannot account for the bird's very rapid decline in its last decades. Elimination of the forests it required at all seasons was an important factor, and the pace of its decline in each region parallels deforestation. Also, diseases affecting the birds or their food trees may have occurred, and like many social species they may have required the stimulation of large numbers for successful breeding.

The now extinct Labrador Duck, which was confined to the northeast coast of North America.

The Carolina Parakeet was the northernmost representative of the essentially tropical parrot family. In the nineteenth century it was still widespread from southern Virginia south through Florida and west to river valley forests in Texas, Oklahoma, and Nebraska. The birds wandered north to central New York, the Great Lakes, and Wisconsin. The last wild specimen was taken in Florida in 1901, and there are possibly correct sight records as late as 1938. The Carolina Parakeet's lack of suspicion made it an easy victim to the farmers who shot the flocks that visited orchards and grainfields. Others were shot as game or trapped to be sold as pets. Habitat destruction was not a significant factor, since many of the large southern forests and cypress swamps they inhabited remained long after they disappeared. Aside from crops, their usual foods were the seeds of thistles, cypress, maple, and elm, all still abundant — hunting seems to be the sole cause of this bird's extinction.

Less well understood are reasons for the disappearance of the Labrador Duck. It may never have been a common species, and, because it was not good to eat, was not shot heavily in its regular wintering areas, the Bay of Fundy, Long Island, and the New Jersey coast. Its breeding range is not known, but may have included Labrador. The large number of thin plates on the sides of the bill suggests a specialized feeding technique, and this, along with hunting or egg collecting by natives in its breeding area, may have contributed to its extinction. The last known Labrador Duck was shot on Long Island in 1875.

A truly extinct bird can never be brought back to life, and the conspicuous habits or narrow range of the birds just noted make us sure they are gone forever. However, there are other birds once believed extinct that have somehow persisted. The Eskimo Curlew was once so abundant that its large, unwary flocks were shot by the wagonload in the Mississippi Valley as they went north to Arctic breeding areas, and on the Atlantic coast as they went south. The birds declined rapidly after 1875. By the early twentieth century, many thought the Eskimo Curlew extinct, but occasional reliable sight records or collected specimens are reported: one was illegally shot in Barbados, West Indies in September 1964 and two birds were seen together on the shore of James Bay, Canada, in August 1976. Since the curlew nests in undisturbed sections of the high Arctic and shooting is already illegal, there is little more we can do to protect the tiny population that survives.

The Ivory-billed Woodpecker shared southern river-bottom forests with the Carolina Parakeet. It fed on wood-boring insects that infest dead or dying trees of a certain age, and therefore required large tracts of mature forest. As these forests were cut, the birds disappeared. The last well-known population existed in a Louisiana forest that was cut in the 1940s. Occasional reports, some possibly correct, still come from searchers in the remaining swampy forests of South Carolina, Florida, Louisiana, and the Big Thicket of east Texas. All likely habitat should be protected, since the ivory-bill, if it still exists, is highly mobile in seeking new feeding areas and may move to one of the forests where none are currently found.

The Bermuda Cahow, a small shearwater that nested only on Bermuda, was thought for nearly three hundred years to be extinct, until it was rediscovered in 1916. It had been abundant when Bermuda was first settled, but during a famine in the winter of 1614–15, vast numbers were taken to keep the colonists alive. The population was so reduced that in 1616 a law protecting the Cahow — one of the world's first conservation laws — was enacted. (The first American bird conservation laws were a closed season on heath hen, grouse, quail, and turkey in certain New York counties in 1708 and a 1710 Massachusetts prohibition on the uses of camouflaged canoes, or boats equipped with sails, in pursuit of waterfowl.) Few Cahows were seen in the next several years, and the bird was presumed extinct. Since its recent rediscovery, the tiny population has been carefully watched on the few rat-free islets where it nests. These rocky islets are not the preferred habitat of the Cahow, which in the past dug a burrow out of softer soil. On the islets it competes for nesting crevices with the more aggressive Yellow-billed Tropicbird. The Cahow survives only because David Wingate, a Bermuda ornithologist, has placed narrow baffles too small for tropicbirds at the entrance to

Cahow nests; this program requires constant monitoring and may slowly increase the Cahow population, now just over 100, but reproductive failure from pesticide contamination remains a threat.

The decline of the California Condor has been watched for more than one hundred years. When the west coast was being settled, condors ranged from the Columbia River in Oregon to Baja California and Arizona. These big birds made tempting targets, and many others died by eating poisoned meat set out by ranchers for coyotes. The current population (about fifty) is centered in a refuge in southern California, a region where open, undisturbed space is disappearing rapidly and the ranches that provided carrion are now being turned into fruit farms. In view of its very low reproductive rate — one egg every two years — and as many birds are believed too old to breed at all, it is doubtful how long the California Condor will survive. A new program supplying deer and goat carcasses may be helpful, but the shy condors may be disturbed by tourists, bird watchers, oil drilling, and other activities within their refuge. A program of captive breeding at the San Diego Zoo has been discussed, but how offspring could be restocked in the wild is an unresolved problem.

The Whooping Crane is the most publicized of endangered North American birds. While never common, different Whooping Crane populations bred in several prairie states, Canada, and Louisiana until about 1900. In winter they concentrated in Gulf coast marshes and the plateau of Tamaulipas, Mexico. Through shooting, marsh drainage, and disturbance, the Whooping Crane by 1953 was reduced to one group of 21 individuals that wintered at the Aransas Wildlife Refuge in Texas and bred at Wood Buffalo Park in northwestern Canada. The population is slowly rising; in November 1976, 69 were counted in Texas. The United States and Canadian governments have several programs to increase the crane population. Since the cranes rarely raise both young if they lay two eggs, second eggs have been taken to the Endangered Wildlife Research Program at the Patuxent Wildlife Research Center in Maryland, where they are hatched and raised with the intention that ultimately their eggs or young will be restocked in the wild. In 1975 Patuxent had 21 Whooping Cranes, but nearly all were still less than six years old, when breeding normally first takes place; so far, none have bred successfully. Another project is designed to spread the crane population, to avoid their elimination should some disaster overtake the entire flock. In 1975 14 Whooping Crane eggs were taken from nests in Canada and placed under Sandhill Cranes in Grays Lake National Wildlife Refuge, Idaho. Nine of the eggs hatched and eight young survived to travel with their foster parents to Bosque-del-Apache National Wildlife Refuge in New Mexico for the winter. It will, of course, be at least six years before this project's success can be judged; the young whoopers may have been imprinted on Sandhill Cranes and thus may not breed among themselves.

It should be clear that the main threats to wildlife, endangered or common, are shooting, man's pests, and habitat loss. Today the massive shooting that destroyed the Eskimo Curlew and Passenger Pigeon is impossible in much of the world, and there are few places left that rats and other pests have not yet reached. Shorebirds and other previously over-hunted birds are recovering some of their former numbers, but will always need protection. Wherever rabbits, goats, cats, rats, mongooses, monkeys, or other introduced animals destroy rare birds or their habitat, an attempt should be made to eliminate them. For the larger problem of habitat preservation, there is no easy solution in either the most conservation-minded nations or in the least. In the Pacific Northwest, for example, the Spotted Owl is now considered rare. It inhabits forests of mature timber, and each pair requires several hundred acres of land whose lumber was worth $1,600,000 per hundred acres in 1975. Is one pair of owls worth $10,000,000? The Puerto Rican Whip-poor-will, a full species rediscovered in 1961 after 50 years of presumed extinction, is confined to three forests totaling 8000 acres. Recent plans for a jetport and a copper smelting plant nearby have been stopped, but 80 percent of the population lives in a forest that has been proposed as a sanitary landfill site.

Many countries have or are developing systems of national parks and refuges, but they are not always adequately protected. The pressures of an ever-increasing world population require the conversion of more land each year to human-oriented uses. Unfortunately, much of the conversion is done in a shortsighted manner that leaves land unusable by wildlife and, in a few years, unusable for people. The destruction of tropical forests is a particular problem — for birds because many forest species are highly specialized or sedentary, and for people because, once cleared, the soil can nourish food crops for only a few years, after which the land is abandoned, never to revert to forest, while new areas are cleared.

Even more difficult than the problems of land conservation are those of chemical pollution. Pesticides like DDT, DDE, and dieldrin, now found throughout the environment and persistent for years after their application, affect birds by interfering with the enzyme-hormone balance that controls calcium production for eggshells, so that heavily contaminated birds lay eggs with shells so thin they break when the bird sits on them; absorbed in large doses, pesticides can kill birds outright.

The birds that suffer most from pesticide contamination are those at the top of food chains in which pesticides are increasingly concentrated. Eaters of large fish and of birds that in turn have fed on contaminated fish or insects suffer more than rodent eaters. The Brown Pelican, Osprey, Bald Eagle, and Peregrine Falcon are among those most severely affected; special conservation programs have been developed to help each of them, but their success depends on an end to poisonous chemicals in the environment. In Louisiana, a successful program of restocking Brown Pelicans, the state bird, which had been completely eliminated, was undone in 1975 when heavy doses of chemicals applied upriver came down the Mississippi. Nearly all Louisiana's 450 Brown Pelicans died, and were found to have endrin, dieldrin, toxaphene, DDE, BHC (benzene hexachloride), HCB (chlorobenzene), PCB (polychlorinated biphenyls), and heptachlor epoxide in their brain tissue. The doses of endrin alone would have been lethal.

Several innovative programs have been developed to help the raptors particularly affected by pesticides. In certain areas where Ospreys and Bald Eagles, for example, have had little reproductive success, eggs from less contaminated regions have been put in the nests; the birds accept these as their own and raise the young. Such programs are a useful way to maintain raptor populations in some places until those areas can again produce their own.

The problem is more complex for the Peregrine Falcon: since its eastern North American population is virtually extinct, there are no nests in which to put eggs from other regions; moreover, nearly every other Peregrine population has also decreased. In the last few years Peregrines have been successfully raised in captivity, and some of the young have been released at former nest sites, but the young falcons must be trained by man to hunt, since they have no wild parents to teach them. It is hoped that when these birds reach breeding age they will return to the release sites; where they will spend the intervening years is unknown — if they go to South America, the traditional wintering area for eastern Peregrines, they will receive heavy doses of the pesticides now banned in the United States.

Specialized problems hamper the survival of certain birds. Golden Eagles and other large raptors in the West die after eating carcasses poisoned for coyotes, are electrocuted when they perch on live wires, and are illegally shot for sport from airplanes. Two to three million waterfowl die each year from lead poisoning, which they get by swallowing spent bullets mixed with vegetation. Thousands of waterfowl and seabirds are killed by oil spills in oceans, lakes, and rivers. The Danish salmon gillnet fishery off West Greenland accidentally kills about one-half million Thick-billed Murres annually; hunting of murres in Greenland takes another three-quarter million birds — combined with other mortality factors, these losses are greater than the annual production of 1.5 million. In the north Pacific, the Japanese salmon gillnet industry takes between 214,500 and 715,000 birds per year (incomplete data account for the differing estimates), including murres (58 percent), shearwaters (27 percent), puffins (9 percent), and Northern Fulmars (5 percent). Whether the annual production of each species involved can sustain these losses is unknown.

Skins, pets, feathers, trophies, and other uses of birds as status symbols make conservation of some species difficult. By the early twentieth century, egrets were nearly eliminated from the United States because their plumes were in such demand for hats; protest, protection, legislation, and a change in fashion protected the last surviving colonies and today the birds are again widespread. Such success stories may not be repeatable for slow reproducers like the Monkey-eating Eagle of the Philippines or the Harpy Eagle of Central America, which are taken for skins, pets, and trophies, and already suffer from deforestation. In Southeast Asia, hornbills are pursued because their bills, ground up, are believed to be an aphrodisiac. Parrots, especially the macaws, are overtrapped for the pet trade, and a high percentage of all caged birds die before reaching their destination — in Costa Rica even the Lesser Goldfinch is seriously endangered by the pet trade.

The Hawaiian Islands have lost more of their native birds than any other comparable region. Sixty percent of the 68 known forms of Hawaiian land birds are either extinct, probably extinct, or rare

The Nukupuu, a Hawaiian honeycreeper extinct on Oahu and Hawaii, may still survive in very small numbers on Kauai or Maui. According to one description, it fed by hammering at rotten bark and wood with its lower mandible and used the upper mandible to reach into the crevices it had created.

enough to be in danger of extinction. Unique to this island group are the Hawaiian honeycreepers, a family of small, colorful passerines, some with long, decurved bills adapted for flower probing. Although forests have been cut, and man's usual pests and a variety of competing songbirds introduced, these factors alone do not explain the rapid disappearance of some Hawaiian birds. Honeycreepers especially are now confined to areas above 3000 feet, even where suitable habitat formerly occupied exists below. The cause of the honeycreepers' decrease has recently been discovered: a mosquito-borne avian malaria to which the birds have no immunity. No mosquitoes had existed on the Hawaiian Islands until 1826, when a ship drained water barrels containing larvae of the Mexican Night Mosquito, a principal carrier of avian malaria, into a stream on Maui. The honeycreepers, having evolved in a mosquito-free environment, did not crouch while sleeping or bury their faces under the wing feathers as do birds of mosquitoed regions, and thus were easy victims to the insects, which bit the featherless areas; honeycreepers quickly died out wherever the mosquitoes occurred. As the mosquitoes cannot tolerate conditions above 3000 feet, birds above that altitude were safe — recent experiments bringing honeycreepers to lower elevations have shown that they still die a few days after being bitten, while individuals kept in the same place in mosquito-proof cages survived.

Finally, there are birds whose extinction we cannot explain at all. The Bachman's Warbler, perhaps never really common, may become extinct in the next few years, although man seems to have played no role. It breeds in heavily wooded swamps and bottomlands, nesting in tangled brush, and in the past extended from North Carolina, south Indiana, and southeast Missouri south to South Carolina, central Alabama, and northeast Arkansas. Many places where it was known to breed in 1900 are still unaltered, as is much of the wintering area in Cuba and Isle of Pines, Bahamas. In 1889, up to 25 or 30 could be seen in one day during fall migration in Key West, but by 1928 it was considered rare, and few were found in traditional breeding localities. Despite extensive searches, fewer and fewer have been seen in the last few decades. One male was seen on the Louisiana coast in 1973 and another male in South Carolina in 1975. No females have been seen for several years, nor have any at all been reported from Cuba or the Bahamas.

People often ask what they can do when the conservation picture looks so bleak. In fact, most conservation causes are won by citizen action. The first responsibility is to know your local areas that shelter wildlife. Study them before they are threatened, and work to have particularly vulnerable areas preserved; state conservation agencies or national environmental organizations may be able to advise you or work with you for their protection.

The Federal government is responsible for vast tracts of ecologically valuable land. Even those areas specially set aside as refuges may not be well managed when budgets are small. Other areas are often leased for grazing, timber, or mineral rights, practices that sometimes conflict with wildlife preservation. Let elected and appointed officials responsible for these matters know when you feel wildlife is not being adequately safeguarded.

During the last few years of increased ecological awareness, a host of new Federal, state, and local laws designed to protect the environment have been passed — watch that they are not watered down, and work to have them strengthened where necessary.

Many local and national organizations devote time, money, and expertise to conservation problems in which the citizen cannot effectively participate on his own. They deserve your support. Some of the ones particularly involved on a national and international basis in preservation of critical habitat and studies of endangered birds are the National Audubon Society, World Wildlife Fund, and International Council for Bird Preservation. Their addresses are listed in the Appendix on page 291.

Since a disproportionate amount of habitat destruction all around the world takes place to satisfy the material wants of Americans, we can make the effort to change our "needs." Millions of acres of land might be saved from mining and lumbering if more metals and paper were recycled. Disasters resulting from drilling for and transporting oil could be reduced if unnecessary petrochemical products were replaced by those made from renewable resources. While we can have little influence over the land-use policies of nations now repeating our mistakes, there is no one who in his personal life can't make some contribution to improving the environment shared by man and wildlife. People interested in birds have traditionally been at the forefront of the conservation movement; careful, knowledgeable observers of wildlife will always be able to make greater contributions towards its protection.

Chapter Fourteen

Attracting and Caring for Birds

ATTRACTING BIRDS to the home or garden has become a popular pastime throughout America. By providing appropriate plantings, food, and shelter, you can significantly increase the number of birds visiting you. You will be amazed at the results, whether you live in the open countryside where you feel birds are already as abundant as possible, in a suburb, or in the city with a small yard or just an apartment terrace that you thought no birds would ever visit. (One man living in midtown Manhattan has seen 99 different species of birds in his yard. He feels that in recent years the number of migrants has decreased, but there are more nesters.)

Bringing birds closer to you gives marvelous opportunities to study nearly all aspects of their behavior. Courtship, song cycle, nest construction, and care of young are some of the things you can observe closely if you provide birdhouses or suitable nesting sites. Feeding, drinking, and bathing preferences, and interactions between birds of the same or different species can be seen at a feeder or birdbath. While watching birds carefully at their routine activities, you never know when you are going to see something that hasn't been recorded before — much of what we do know about bird behavior comes from this sort of patient observation. Attracting birds is valuable in material ways as well: increasing the number of insect eaters present in summer may contribute directly to your comfort, and may aid the health and looks of your garden. It should also save you the use of some pesticides.

Feeding

The most "natural" way to feed birds is to provide the appropriate food-bearing plants. This is simpler than stocking a feeder, since the plants after a short time will require little care, producing food independently of your presence, budget, or shopping schedule. If you are starting or maintaining a garden, you will need trees and shrubs anyway, and many plants valuable to birds have horticultural interest as well. In planning a bird garden, consider which seasons you visit the garden and which birds are there to be attracted. If, for example, your garden is at a summer house, look for shrubs or vines that bear fruit then, so that you can witness the results of your labor. If yours is a year-round garden, choose plants that provide food during every season. Various trees and shrubs fruit in spring, summer, and fall, and the fruit on others lingers through the winter and into the following spring. A well-balanced garden can feed birds all through the year.

Your garden should have an expanse of lawn; this provides ground feeders with worms and grubs and you with a viewing area. Thick bushes where birds can nest and take shelter are essential, and these bushes may also be food-bearing. Where possible, don't overprune, or remove dead wood; dead branches or whole trees are especially attractive to woodpeckers, and a pair may nest there and provide a hole that will be used year after year by other birds. The flowers of some trees seem especially attractive to insects, and these in turn attract birds. In the northeast, several of the oaks that flower in mid-May, when the bulk of the warblers are passing through, hold many more migrants than do surrounding trees; flowers of the native black cherry (*Prunus serotina*) also lure warblers and vireos. In selecting new trees for your garden, it may be worth spending a spring watching to see if any varieties in your area have this effect.

The garden flowers you may have thought purely ornamental can also feed birds if you let them go to seed. Chrysanthemums, marigolds, petunias, snapdragons, sunflowers, and zinnias are some of the popular garden flowers with seeds attractive to birds. Almost any flower is of some interest to a hummingbird, but an abundance of tubular red flowers may lure them to linger or nest near you. Many hummingbirds specialize in flowers too deep to be pollinated by bees or other insects; some wildflowers now depend exclusively on hummingbirds for pollination. In the West, the migration of hummingbirds and opening of certain flowers is synchronized. The large number of western hummingbirds and the variety of their habitats preclude a listing of suitable flowers here; *Hummingbirds and their Flowers,* by Grant and Grant (Columbia University Press, 1968) has a list several pages long of flowers particularly attractive to western hummingbirds. In the East, the Ruby-throated Hummingbird favors, among others, trumpet vine (*Campsis radicans*), trumpet honeysuckle (*Lonicera sempervirens*), beauty bush (*Kolkwitzia amabilis*), butterfly bush (*Buddleia* sp.), weigela (*Weigela* sp.), foxglove (*Digitalis* sp.), and snapdragon (*Antirrhinum* sp.).

If you are planting new shrubs, vines, or trees, it is wise to check with a horticultural authority as close to home as possible, since no book specifying bird-attracting plants can describe what is best for every type of soil, climate, amount of sun or shade, or other local variable. Chapters of the National Audubon Society, garden clubs, nurseries, wildlife refuges, or nature centers are good sources of advice, particularly in the West, where conditions vary so much from one locality to another and where there are still fewer published authorities.

In the East, tartarian honeysuckle (*Lonicera tatarica*) puts out fruit in spring. In June, ripe mulberries (*Morus* sp.) attract robins, catbirds, orioles, and many others with their fledged young. By July, chokecherries (*Prunus virginiana*), highbush blueberry (*Vaccinium corymbosum*), and wild blackberry (*Rubus alleghaniensis*) are fruiting. From late summer into at least September, other cherries (*Prunus* sp.) and grapes (*Vitis* sp.) have fruit. Pokeberry (*Phytolacca americana*) is a tall weed with deep purple fruit favored by thrushes in September. Off in a corner, it is rather pretty; the birds spread its seeds and it is difficult to eradicate. Firethorn (*Pyracantha* sp.) produces many berries in October, when hawthorns (*Crataegus* sp.) and viburnums (*Viburnum* sp.) also come into fruit. The latter two retain their fruit into winter. Other popular plants with particularly long-lasting fruit — from fall often into the following spring — are autumn olive (*Elaeagnus umbellata*), bayberry (*Myrica* sp.), Japanese barberry (*Berberis thunbergi*) and multiflora rose (*Rosa multiflora*). Some of the pulpier fruits will do double service — the outer part will be eaten by thrushes, robins, and Mockingbirds, and the seeds, dropped or passed through their system, will be taken by sparrows or chickadees. Bird-gardening and horticultural books will give you a much expanded list of useful plants, with details on their cultivation.

Insecticides are of course to be avoided in any garden designed to support birds. Among the worst are ones containing dieldrin, endrin, aldrin, toxaphene, and heptachlor. Brand names do not indicate an insecticide's ingredients: you must read the small print. Certain herbicides are also dangerous to birds: these include 2,4-D and 2,4,5-T. A few insecticides are available that do not affect warm-blooded creatures, but to avoid killing bees with them, spray in the evening. Pyrethrum, rotenone, nicotine, malathion, and derris are the safest. Derris is dangerous to fish, so do not use it where it may reach a steam or pond.

Naturally, not all the birds that pass through your garden will be lured to stay by your plantings, especially in winter. Some birds never eat seeds or fruit, and others require a more varied diet. To attract some of these birds, there are many kinds of food you can put out. *Never* undertake a feeding program, however, unless you are positive you will be able to continue it. In winter particularly, it is essential that the food you supply remain constant without fluctuating. By supplementing the food naturally available, you are supporting a larger than natural population at a higher density around your garden. If their food supply is suddenly cut off for even a few days while you take a winter vacation or can't get to the store, they may starve. Similarly, if certain species take only certain foods from your feeder, don't expect them to switch if you run out of the usual. In the northern states, to attract southbound birds to stay at your feeder, you should begin stocking it in October, and should continue until at least mid-April, or whenever the last irregular cold periods end in your area. Some people tame their local Cardinals, chickadees, or titmice to the point where they will take food from the hand. While some birds thus tamed clearly recognize their usual feeders, others will incautiously approach or even land on almost anyone, especially when hungry. To keep your bird guests from later landing on someone less sympathetic, let them stay wild.

Feeding can be maintained the year round, but you may want to decrease or change the types of food during the warm seasons, when birds different from the winter population will be present. In summer, your special plantings should be sufficient; in any case, this is the time you want birds to be eating insects. During each season you maintain a feeder, be regular about what you supply and the hour you supply it. Ideally, feeders should be filled early in the morning, especially in winter, because birds lose the most weight over cold nights.

A feeder should be placed in the open, but not where it receives too much sun or direct wind. Thick bushes several feet off give birds a place to take refuge, but should be far enough away to prevent a cat from making a concealed approach.

To keep squirrels from reaching a tray-type feeder, place it on a metal pole, or put a metal sleeve or cone, curved downward and at least 18 inches in diameter, on the pole; do not put the feeder directly under a tree.

The simplest type of feeder is a tray or table, set on a pole about 5 or 6 feet off the ground. A 3- or 4-square foot tray is more effective than many of the smaller ones sold. Its sides should have rims, to prevent the food blowing off; a break at the corners of the rim makes cleaning easier. A hopper is most efficient if the tray serves seeds. A sloping roof a foot over the tray makes it more inviting in rain or

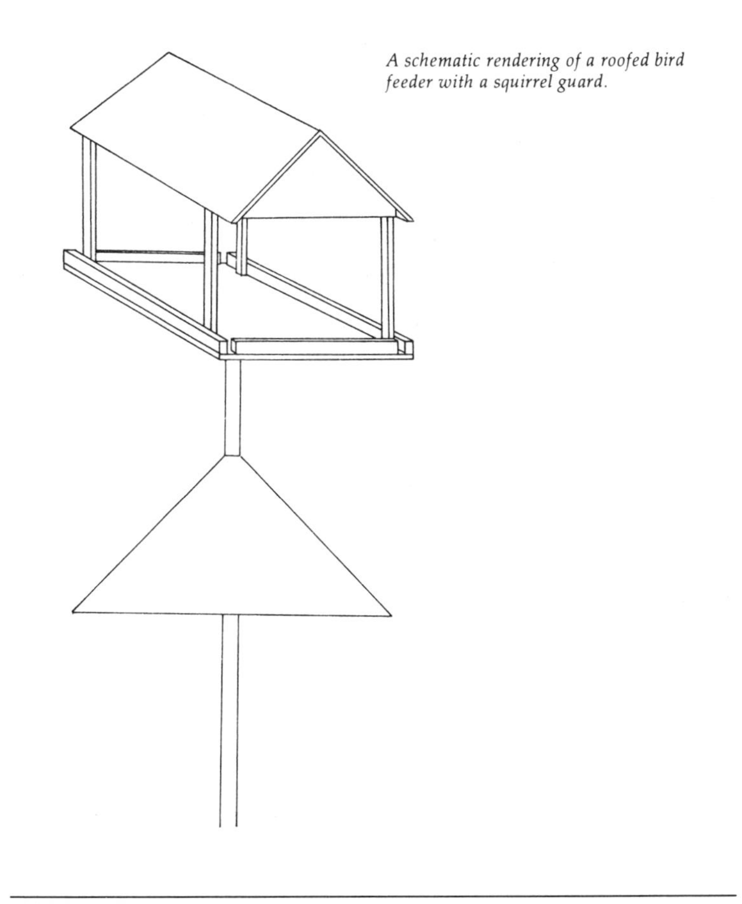

A schematic rendering of a roofed bird feeder with a squirrel guard.

snow. Similar trays can be suspended from a branch (better where there are no squirrels), from a stretched wire, or a window ledge. The many books on bird feeding will give you specifications and other designs.

When starting a feeder you might use a compartmentalized tray, with different foods in each section to see which are popular. There are many commercial birdseed mixtures available. You can also provide dried berries, fruit slices, bread crumbs, millet, buckwheat, chick cracked corn, peanuts, sunflower seeds, or other vegetarian items on your tray. Many birds are fond of peanut butter — it should be mixed well with equal parts of cornmeal to keep it from caking the mandibles; the mixture can be placed in one of the feeders specially designed for it, or smeared on the rough bark of trees.

Juncos, White-throated Sparrows, and other ground feeders rarely if ever come to a tray off the ground. Seed for them should be scattered in an open, windless place several feet from shrubbery. For insect eaters like woodpeckers, nuthatches, and Brown Creepers, suet held in a wire mesh against a tree trunk is very attractive; be sure it is protected from rain and sun. Almost any sort of table scrap, served in the appropriate size and place, will interest some bird, and anything they don't like, they simply won't take. If jays, grackles, or Starlings seem to be monopolizing your feeder, they can often be lured away by a supply of scraps placed elsewhere. The bird-feeding books go into much greater detail on food mixes, feeder styles, and ways to discourage common pigeons, Starlings, and House Sparrows.

Feeding hummingbirds is strictly a summer pleasure in most of North America. Many commercial feeders are now available, but you can make your own from three-inch-long test tubes suspended from a wire. A red plastic flower or a bit of red ribbon attached to the tube seems to catch their attention. Fill the feeder with a mixture of two parts water and one part sugar, melting the sugar into the water, or three parts water to one part honey. Dissolve the honey in hot water and boil briefly to prevent fermentation.

Drinking and Bathing

Birds require water as well as food throughout the year. If there is no natural source of water on your property, you can create a drinking place and birdbath, but the same cautionary remarks about feed-

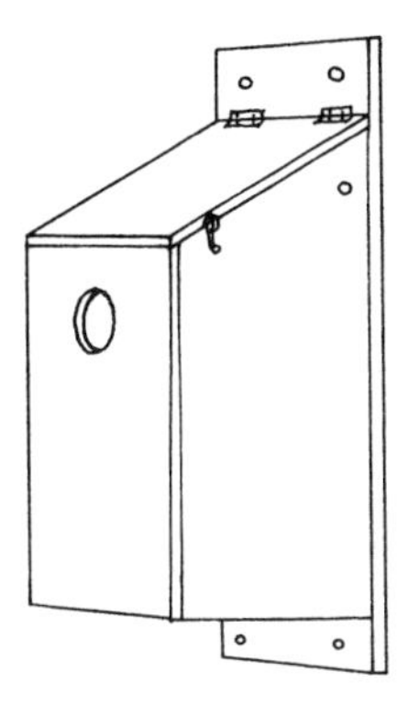

A schematic rendering of a bird house to be attached to a tree for small hole-nesters.

ers apply to water as well, particularly in winter when water in the birdbath is likely to freeze. Several anti-freezing devices are now available. The bath should have an unslippery surface and be very shallow, at least at one end. It may be on the ground or raised, but should always be placed in the sun, several feet away from shrubbery: birds are least wary while bathing.

A trickle or drip of water is very attractive to birds. If the bath is under a tree branch, a filled bucket with a tiny hole in the bottom or a suspended garden hose adjusted to the finest drip will draw the attention of more birds. A bath in the open can still be reached by a hose running along the ground. Always be sure the ground around the bath is well drained.

Many small birds also bathe in dust, possibly to reduce parasites. A few square feet of powdery sand, earth, and ash, a few inches deep, in a sunny place, will attract several species.

Shelter

Most of your plantings will provide shelter as well as food. Dense bushes, vine-covered walls, or thorny tangles are nesting sites for many birds and winter roosting places for others. Evergreens are particularly attractive in winter; some that offer both food and shelter are yew (*Taxus baccata*), Canadian hemlock (*Tsuga canadensis*), white pine (*Pinus strobus*), and red cedar (*Juniperus virginiana*).

Providing birds with nesting sites is an old tradition. The Romans built houses for pigeons, and in the southeast United States the Indians hung hollow gourds for Purple Martins. Gourds are still popular martin houses in the south. As noted in Chapter 8, the breeding density of many hole-nesting passerines seems to be limited by the availability of holes, so placing appropriate bird houses can significantly increase a local population.

Specifications for size and placement vary with different species; you can house hole-nesters all the way from chickadees to Screech Owls to Wood Ducks. Many well designed bird houses are available; books on attracting birds also give you full details on how to build your own. In general, boxes attached to a tree trunk should be several feet off the ground, not facing into prevailing spring winds, and with front angled downward so that rain does not enter the hole. In the wild, southerly entrances are most frequent. The site should not be surrounded by thick foliage, as parents need an open flight path to and from the nest. Don't put a perch on the nest box; this makes it easier for cats, squirrels, and other predators to reach in. The roof or one wall should be hinged, so the box can be opened and cleaned after use, and there should be small drainage holes in the bottom. If you put a box on a pole, the same anti-squirrel devices used for feeders should be included.

Ideally, bird houses should be put up in the fall. This gives you a chance to note and correct any structural flaws like leaks or warping. During the winter the houses may be used as roosting holes and will be inspected by the permanent residents. Boxes for summer residents should be placed at least shortly before their expected arrival in spring.

A hole 1⅛ inch in diameter will exclude Starlings and House Sparrows, which often monopolize nest boxes and even evict more desirable species. Bluebirds require a 1½ inch hole; if a box with this size hole is placed on a fence post in open fields or on a pole only 3 or 4 feet off the ground, it will not attract Starlings or House Sparrows.

Plastic milk jugs have one environmentally redeeming feature in that they can be used as birdhouses, suspended from trees like gourds. Paint the jug black to reduce the light inside, then white to reduce the heat, cut the appropriate size hole in the side and a few drainage holes in the bottom, and hang from a relatively short rope so that it doesn't swing too much. More widely holed jugs, without the drainage holes, may work as feeders.

Several birds that do not nest in holes, such as Eastern Phoebes and American Robins, will use a tray-like ledge with a roof over it, attached to a tree or the side of a building. Barn Swallows will use an unroofed shelf under the eave or inside a building. Wooden strips nailed under the eave on the outside of a barn or other large building in the open will make it easier for Cliff Swallows to attach their nests. After the nesting season is over, knock off the swallow nests, so that each pair will build a new one — this avoids occupation by House Sparrows before the swallows return. Purple Martins nest colonially in martin "apartment houses," that can be purchased or made according to specifications. The gourds used for martins in the south are Lagenarias, or "bottle gourds," several suspended from crossbars on a tall pole. Ospreys, relatively fearless where they still nest, sometimes build their nest on an "Osprey pole," a stout telephone pole about 30 feet tall, with a platform or crossbars six feet in diameter placed at the top.

In city parks, robins and grackles often incorporate cast-off paper, plastic wrappers, and tissues in their nest. You can provide many birds with more attractive nest material. House Wrens filling their nest hole with twigs will come to your lawn to pick twigs from a pile you provide. Barn Swallows will swoop down to pick up white feathers for nest lining. Yarn, string, or shaggy dog hair are suitable weaving materials for many birds; never offer pieces more than 6 inches long, since a bird may become entangled or strangle itself when using a longer piece. During dry springs especially, a patch of mud will make nest building easier for robins, grackles, and Barn and Cliff Swallows; it is particularly interesting to observe the swallows, which never come to the ground except when gathering mud.

Caring for Birds

Every spring and summer, people find fledgling birds looking rather helpless on the ground or in a bush, and "rescue" them by taking them home. Usually the parents are nearby, although not necessarily visible, and the young bird should be left alone. If someone brings you a feathered young bird, take it back to the exact place it was found as soon as possible (putting it on the territory of another pair of the same species will do no good). If it is not fed in the next several hours, then you may consider it abandoned. There is no truth to the belief that parents will forsake a young bird once it has been touched by humans.

A young *un*feathered bird out of a nest has probably fallen, and cannot be far from home. Look carefully for the nest and put the bird back in if possible. Sometimes a nest on an open branch, like a robin's, will blow over. If the young are alive and the nest in good condition, replace them in a securer location visible to the parents. If the nest is destroyed, a small fruit box or cardboard box lined with soft material is an adequate substitute.

If your young bird seems definitely abandoned, you may attempt to raise it, but this is a time-consuming job. Place the bird in a warm, softly lined box. The box should be kept dry, clean, and out of wind and direct sunlight. A light bulb near the box can provide warmth. The young bird should be fed every twenty minutes, once an hour at the minimum, from — more or less — dawn to dusk, as it would be by its parents. A mixture of finely chopped egg yolk and minced canned dog food or hamburger meat, held together with raw egg white or milk, is often successful. Bits of soft fruit or berries can also be fed to older birds. There are many variations on this recipe. Don't ever give a young bird liquids — they may make it drown; your food should have all the moisture required. Young birds will usually open their gape and call if you approach when they are hungry. Using a toothpick, a narrow wooden spoon, or a thin paintbrush, drop bits of food far down the throat: young birds cannot swallow food simply placed in their bill. Fledglings should be in a cage, outside in good weather, with food on the ground or the floor. When the bird reaches adult size, let it out of the cage to find food you have put on the ground, then gradually decrease the amount of food you provide so that it learns to find natural food; eventually the bird will attain complete independence.

A sick or injured adult bird is more difficult to treat. Occasionally you find dazed birds that may just have hit a window and need only to rest in a quiet, warm place for a few hours. Give any bird a warm, dry box lined with newspaper and containing a perch off the ground. A shallow plate of water and one of food similar to what has been described (unless you are familiar with the species' diet from having seen it at your feeder) may be all it needs.

A bird that is bleeding or has a broken bone should be taken to a veterinarian. If this is impossible, find the source of the wound, clean it, disinfect it, and if necessary sew it. A broken leg should be disinfected, and set with a splint (toothpick or match stick) and tape. A broken wing should be held immobile with a wrapping of stocking or cellophane tape and disinfected at the break with sulfa dust. After a few days the wing should be freed for exercise and preening. Do not expect a high success rate for adult or young birds in your care — remember that mortality is always high in natural situations. Your main goal should be to prevent possibilities of injury or death resulting from housing and feeding efforts that have increased the birds' dependence on you.

Aside from the pure pleasure of having more birds around you, attracting birds to your home, where you can observe them closely and often, provides unrivaled opportunities to learn the details of their daily lives. Similarly, few studies of wild birds could give you a better understanding of their bodily requirements and resiliency than can caring for a young or injured bird. The educational and research projects provided by feeders or birdhouses are unlimited — the most worthwhile ones of course are those using data gathered over several years, but even the most casual investigations will repay your effort with insights into the daily needs of the birds around you.

Chapter Fifteen

The Golden-backed Mountain-Tanager, discovered in Peru in 1973 (from a painting by John P. O'Neill, WILSON BULLETIN, *76, 1974.).*

Ornithology Today

MANY OF OUR earlier chapters have ended with speculation. There is much we do not know about birds; in some cases we do not even know how to investigate what we wish to know. However, answers to some of the questions previously thought unsolvable are coming, through the application of new research techniques or the integration of usually separate disciplines. Ornithology's fundamental concern with the evolution and function of the living bird ties together work in physiology, behavior, ecology, and systematics. Even the most anatomical, museum-oriented, or mathematical studies are intended to explain an aspect of how a bird operates in its environment. The following review of some current major topics in ornithology will not be complete, but should give you an idea of both the breadth and unifying elements of ornithological research today.

Inventorying

Even with the world rather thoroughly explored, there are still bird species unknown to science: between 1970 and 1976, at least 25 new species were described. At the time of this writing, in mid-1976, publication of descriptions of several other species — 6 from Peru alone — is pending; more species undoubtedly remain to be discovered. Most of the new species come from remote parts of the world, or rugged, inaccessible habitats within previously explored regions. Among the recently described discoveries are a grebe in Argentina; a swift from the Cook Islands in the South Pacific; four Brazilian hummingbirds; a kingfisher in the Cook Islands; two asiatic bulbuls; two larks in Ethiopia; a nuthatch, a thrush, a warbler, two flycatchers, and two weaver finches from the Old World; three Peruvian tanagers; a wood warbler in Puerto Rico and one in Chile; a Hawaiian honeycreeper; and a small passerine of unknown affinities from Peru.

The formal description of a new form, species or subspecies, includes a scientific name, thorough description of the bird in every known plumage, precise location of discovery, range as far as known, and whatever has been learned about its ecology and behavior. The describer attempts to place the form with its closest relatives. Not all the newly discovered birds have English names as yet; some come from non-English-speaking areas and were described by ornithologists writing in other languages.

Many of the new species seem to have very restricted ranges, and little is known about their life histories. The Hooded Grebe is part of the genus *Podiceps,* which includes most North American species. It was discovered in 1974 at Laguna Los Escarchados, a small lake in southern Argentina. The grebes nest there but leave before the lake freezes over in late fall; because it has salt glands, the Hooded Grebe is presumed to winter on salt water, probably on the Atlantic coast, since to reach the Pacific would involve flying over the Andes. A survey in 1976 counted 73 or 75 Hooded Grebes at Laguna Los Escarchados and none on the similar lakes nearby.

The three new, brightly colored Peruvian tanagers all come from high Andean habitat known as "elfin forest" at the edge of the tree line, where trees are short, gnarled, and thickly covered with lichens, orchids, and bromeliads. Tree ferns and bamboo also grow in this habitat, which receives heavy rainfall and is enveloped in thick fog for many daylight hours. The birds have been named Rufous-browed Hemispingus, Golden-backed Mountain-Tanager, and Parodi's Tanager; each has a different ecology and is found in different sectors of forests on the eastern slopes of the Andes: the hemispingus occurs in upper limits of the elfin forest, traveling in mixed species groups and never feeding more than 6 feet off the ground; the mountain-tanager is restricted to scattered islands of elfin forest above the main forest, feeding in pairs in the upper sections of thickly leaved trees; Parodi's Tanager inhabits a mixture of elfin forest and tall grassland.

In the same Peruvian elfin forests a previously unknown little brown and yellow bird was discovered in 1973. Called the Pardusco, the name applied to it by the discoverers' Peruvian field assistants, ornithologists have not found characteristics that would place it positively with any of the tanagers, honeycreepers, or finches, the three most likely groups in the region. The Pardusco lives in the edges between elfin forests and sphagnum bogs at 9800 to 11,500 feet, and feeds on insects in the lower vegetation while traveling in groups of 5 to 15, sometimes with other species.

The Elfin Woods Warbler, discovered in 1971 in Puerto Rico, lives in unbroken forests on the upper slopes of the Sierra de Luquillo, east of San Juan. It somewhat resembles a Black-and-white Warbler and has been placed in the genus *Dendroica,* which contains many familiar North American species. How a bird escaped notice on such a populated island that has been the subject of many ecological surveys can only be explained by the remoteness and impenetrability of the elfin forest and this species' unobtrusive habits. Its population is probably no more than 300 pairs.

In a similarly impenetrable rain forest on the Haleakala Volcano in Maui, a new Hawaiian honeycreeper was discovered in 1973; named Po'o Uli, Hawaiian for "black faced," it is different enough from all other honeycreepers to merit the creation of a new genus, *Melamprosops.* The Po'o Uli has a short, moderately stout bill, feeds exclusively on insects that it picks and pries from cracks in bark, and is probably restricted to the forest in which it was found.

Some birds discovered long ago are still as unfamiliar as the newest discoveries. Known only from a few old specimens in museums, these are birds that were collected once or twice and never seen again; usually nothing is known about their ecology. One such bird, recently rediscovered, is the Imperial Snipe of the Andes. A single specimen was described in 1869, and then for a hundred years the bird remained completely unknown. Recently, ornithologists working in the northern Cordillera Vilcabamba of the Peruvian Andes made a concerted attempt to collect a snipe-like bird they found giving song flights around their camp at dusk. The bird was active only in the hours it was practically impossible to see, and the rugged and steep terrain made pursuit too difficult; by chance, one flying lower than usual landed in a mist net, and proved to be the long-lost Imperial Snipe. Others were heard calling elsewhere along the mountain ridges, but were never seen in the daytime. As the habitat over which they fly is a mosaic of elfin forest and marshy grassland, it is still not known whether the Imperial Snipe lives in the forest like a woodcock or in the grasslands like a typical snipe.

Fossil discoveries are continually increasing the list of birds known to have existed. Bones found recently in Wyoming have been ascribed to a species of owl that lived early in the Eocene, a period beginning 60 million years ago. It may be the earliest owl known, and of a group restricted to North America; bones of owls more closely resembling forms alive today have been found in Europe from the late Eocene. A species of anhinga larger than any other extinct or living forms has been described from bones found in Nebraska; it lived in the Pliocene, a period beginning 12 million years ago, when the Nebraska climate was considerably hotter than today. Recent Pleistocene discoveries, perhaps 1 million years old, from central Mexico include a grebe and a flamingo similar to living forms. In the same locality, bones of Western Grebe, Lesser Scaup, American Coot, and American Avocet were found, giving us a better idea of the age and range of these species. A recently found Pleistocene White-winged Scoter from Maryland is the earliest Atlantic coast record of this species; all other Pleistocene finds have been in California and Oregon.

Systematics

After the discovery of a new bird, the first requirement is to place it with its most closely related forms. As we have seen, systematics is an ongoing discipline, with ideas of the true relationships and evolutionary sequence of many well-known birds still unresolved. The American Ornithologists' Union's Committee on Classification and Nomenclature is continuously evaluating and synthesizing all new research that has any application to systematics. In a few years the committee plans to publish a new check-list that will give the range of every recognized species known to occur in North and Middle America.

Systematists working in museums are responsible for the recognition of some "new" species. A combination of museum and field study led to the naming in 1973 of the Tepui Swift, a distinct population of southern Venezuela and neighboring Guyana and Brazil that was formerly considered part of the Chestnut-collared Swift, known since 1817. Similarly, the Thayer's Gull, Willow Flycatcher, and Great-tailed Grackle from North America are newly recognized species of the 1970s, in addition to the 25 that were totally unknown. At the same time, of course, other forms previously considered full species, such as the Ipswich, Dusky Seaside, and Cape Sable Sparrows, have been treated as isolated races of more widespread species, and populations found to hybridize extensively where they overlap have been put into one species. (A more complete list and discussion is found in Chapter 3.) New subspecies continue to be described by workers studying large series of skins or finding populations in new areas.

Investigations of hybridization are especially important to systematic research, since they indicate under natural conditions what are "good" species and what are better considered long-isolated subspecies. Man's alterations of the North American environment have been responsible for several of the hybridization zones between previously isolated populations. Montana, Nebraska, and the Dakotas, for example, are states where Yellow-shafted Flickers, Baltimore Orioles, and Indigo Buntings have expanded westward with increased tree planting and irrigation, to encounter populations of Red-shafted Flickers, Bullock's Orioles, and Lazuli Buntings adapted to the drier natural habitat. The hybridization rate has been different in each of these species pairs: flickers in the areas of overlap are almost all hybrids, while the oriole and bunting hybridization zones are narrower. There is some evidence that the orioles are now hybridizing less frequently than when the populations first came into contact, and there are still populations of pure Indigo Buntings well within the Lazuli range. An accurate idea of the status of each population will require many more years of study.

The recent discovery in southern Texas of hybrids between Barn and Cave Swallows is particularly interesting to systematists because the two species have been considered related distantly enough to be placed in separate genera. Cave Swallows were formerly restricted to nesting in caves and sinkholes, but have begun using highway culverts and other manmade structures where they encounter Barn Swallows. Their protein structures, often used as an indicator of degree of relationship, were analyzed and found to be nearly identical. However, such cases of hybridization do not prove that the two forms are one species, since, in this case, one would expect first to see many hybrids between Cliff Swallows, clearly close relatives of the Cave Swallow, and Barn Swallows in the wide breeding area they share. Sometimes, in fact, the behavioral or reproductive barriers between species of different genera that are usually isolated are weaker than those within a genus, or where two species have had long contact: in hummingbirds, for example, hybrids between species of different genera are much more common than those within a genus.

All aspects of bird biology are now used for systematic research; traditional approaches, using plumage and anatomy, remain important, but behavior, biochemistry, and vocalizations are considered as well. Sometimes the pooled evidence is entirely corroborative, at other times it is contradictory. Analyses of the structure of egg whites, of blood proteins, and of DNA, the nucleic acids that function in the transference of genetic characteristics, have clarified some relationships.

Vocal differences were one of the first indications that the "Traill's Flycatcher" might contain two specifically distinct species. A recently developed technique is to play recordings of songs and calls to populations under study to measure their reactions, since birds respond to vocalizations of their own species but usually disregard those of others. This technique recently revealed that populations of a *Myiarchus* flycatcher (the genus of the Great Crested and Ash-throated Flycatchers) in Venezuela previously considered one species are actually three distinct species sometimes occurring together — the three look very similar, but each responded only to tapes of its own calls.

Vocalizations can also show degrees of relationship within one species, as it has within the seven North American subspecies of the Pine Grosbeak; those from Newfoundland to Alaska have a location note described as a clear, emphatic whistle, while the three forms on the west coast south of Alaska have a modulated location note. Flocks with differing location notes do not mix when they encounter one another in winter, suggesting that the forms with each call note are more closely related to one another than to those of the other group; the information is valuable because this kind of judgment could not be made on the basis of appearance alone.

Ecology and Life Histories

As has been indicated often, we know practically nothing about the life histories of many birds, and not that much about some North American and European species familiar to ornithologists since the origin of the science. Integrations of behavioral and physiological work are increasingly common in life history studies investigating, for example, how vocalizations are learned and produced, how hormones affect daily activities, the rates at which young birds grow, how birds respond to stimuli in their environment, and other topics we have surveyed in earlier chapters. As new methods of analyzing data are developed, we find new ways to study even the best known birds. Of particular interest today is how birds adapt to stressful environments like deserts and the high Arctic; the ecology of tropical birds, differing, as we have seen, in so many ways from that of the better studied temperate species, is also an important research topic.

Almost any approach may yield valuable information. Traditional observation studies are needed for many birds, while careful examination of anatomical features will reveal adaptations not evident from watching the living bird. Experiments with wild birds including manipulation of the clutch or brood size can indicate factors limiting population or nesting success. A study of Snow Buntings, for example, found that when given more chicks, they were able to raise larger than normal broods. That the clutch size is smaller than the raisable number may indicate Snow Buntings can only lay a limited number of eggs or incubate a smaller clutch.

A fairly recent approach to life history studies has been the investigation of time budgets, daily and annual; this requires observation throughout the hours a bird is active each day and throughout the year. Such work has shown that studies based on observations made the same hours each day and then extrapolating the activities of the other hours were not accurate, because birds spend more time during certain hours feeding themselves or their young, preening, bathing, resting, etc., than during others, and these schedules may vary in different seasons or different individuals. The amount of time a bird devotes to each activity indicates its relative importance to survival. Predictably, for nearly all birds feeding requires more time in some seasons than in others.

Birds use the most appropriate hours for each activity. Sage Grouse, for example, display at leks during the periods when light is faintest, at dawn and dusk. This reduces their vulnerability to predation by the Golden Eagles active in lighter hours and the bobcats and weasels active at night. Hummingbirds in regions where the nights are cold feed as early and late in the day as they can, to avoid depleting energy resources used to maintain a high temperature through the night; courtship activities take place later in the day, when energy demands have been met. During the laying stages, female Gadwalls spend twice as much time feeding as do their mates, while the males use some of this time to defend feeding territories from other pairs; as the nesting season progresses, males have more and more time available for resting.

A seasonal event that has received increased study in relation to other parts of the bird's time budget is molt. Birds, of course, do not allocate particular minutes of the day to molt, but the energy demands of new feather growth usually require molt to take place at that time of year when birds do not have other heavy demands. Water Pipits, for example, go through their basic molt while in their northern breeding areas, after the young have fledged, at the time of maximum food abundance; males begin molt two weeks before females. The Whitethroat, an Old World warbler, begins its molt in the breeding area if food is abundant, stops during migration, and, if molt has not been completed, finishes in the wintering area. The Piñon Jay, with an irregular breeding season dependent on cone crops, generally begins molt in summer, after breeding has usually taken place, but if food resources delay breeding until late summer, the molt will stop. In the Marsh Hawk, molt of primary feathers proceeds independent of the stage of breeding and always begins near a certain date, with males molting later but faster than females. In many tropical birds molt seems to overlap breeding activities, implying that in certain environments the energy demands of both can be met at once.

Investigations of the amounts of energy required for each of a bird's activities are closely related to time-budget studies. Since the uses of time and energy are the most basic units in the life history of any organism, their study encompasses all aspects of its ecology, behavior, and physiology. Supplies of time and energy are important selection pressures that mold adaptive strategies. To measure energy budgets, one must know food intake, its caloric value, and how much weight is lost in each activity; as these are not easily measurable data, precise information is available for only a few species.

Hummingbirds, because their small size makes them expend energy proportionately more rapidly than larger birds, have been of particular interest, and the caloric value of their food can be easily measured, especially in laboratory situations where they are fed an artificial diet. The building of a well insulated nest sheltered from wind, loyalty to particular perches in cold or hot weather, and specialization in nectar-rich flowers can all be explained and quantified in terms of energy requirements. Nectar-feeding African sunbirds seem able to assess the energy production levels of their territories; measurements have shown that they defend an area that supplies precisely their requirements — if nectar production increases within the territory, they save energy by defending a smaller area; if production decreases, they expand territory size.

The flow of energy from one organism to another within an ecological community is another way to look at physiological and behavioral adaptations. This is also an area in which data are difficult to obtain. A simplified example of an energy flow study might be an investigation of the amount of food used by a seabird colony during one breeding season. Such studies have important implications for conservation, since they can tell us, for example, how many tons of fish are required to maintain seabird colonies of various sizes. The amount of food taken per day by the average individual of each species for itself and its young would be calculated, and each species in the colony would be censused, with the length of its breeding season noted; these figures combined give the amount of food taken from the sea, while an investigation of the food items themselves gives their caloric value. Some of these calories are dissipated as energy, others ultimately serve as food for seabird predators, and still others become guano washed back into the sea to nourish organisms that will feed the fish taken by seabirds. Since nutriments produced by the birds maintain the fish population, the effect of reducing the bird population must be considered in calculations predicting future yields. All this information can be expressed numerically in a mathematical model that predicts the amount of energy used at every stage.

How birds utilize the resources available and partition their environment with other species is another way of looking at time and energy budgets. We have seen that many birds can live together because of the restricted way each forages: five kinds of wood warblers, for example, feed in different sections of the same trees without directly competing, many shorebirds take food from one mud flat by probing to different depths, and Common and Lesser Nighthawks coexist in the southwest because the Common feeds only within its territory while the Lesser wanders unrestricted over large areas. Hermit and Swainson's Thrushes are segregated by time where they utilize the same habitat — the Hermit begins breeding several weeks before the Swainson's arrives, so their peak demands for food occur at different times. For other birds, a division of food resources is not as crucial as other habitat requirements. The way nest sites are selected by each species in mixed heron colonies, for example, reveals how well each has adjusted to competition from the others; generally, the largest heron species present nests highest in the vegetation, since when danger arises it needs the most space to get away. One reason for the Cattle Egret's rapid spread is that it disregards the North American nest placement hierarchy, placing its nest at all levels and often enjoying the advantage usually limited to larger species.

Studies of species diversity within various environments are now used to create mathematical models that can predict the diversity of similar environments elsewhere. Diversity studies may sample a single community, a collection of similar habitats within a region, or the total diversity to be found in all habitats in a large geographical area. In measuring the bird species diversity of temperate forests, the number of foliage layers was found to be a more important factor than the type of trees involved or their diversity; in the tropics, however, forests of a single tree species support significantly fewer bird species than do mixed forests; temperate forest models are further inapplicable to the tropics because they contain too few layers. One of the problems in gathering the data needed to formulate tropical models is the height of the trees and the difficulty of assessing the population of species and individuals in the upper layers. A recent comparison of five tropical lowland forests, two in Asia and three in South America, found that each was different in the total number of species present and in abundance levels, but had similar numbers of birds of each size category and each feeding technique.

Some of the other important factors in diversity studies are the presence of non-bird competitors for a particular niche, historical factors affecting present bird distribution, and size of the area studied. A small woodland identical to a larger one will have fewer species because some birds require large territories and others cannot or will not travel to isolated patches of suitable habitat. The implications for conservation are clear: sanctuaries must be large enough to support populations of all species that use the habitat — saving small patches of "ideal" habitat may not help the animal species one is most anxious to protect.

Migration, Orientation, and Distribution

Recent laboratory studies of what environmental cues birds use for navigation during migration have been extremely enlightening, but many more such investigations will have to be made before we can make generalizations, since the initial studies described in Chapter 10 have shown that some celestial navigators require a specific seasonal star pattern while others do not.

Tracking migrants has always been a difficult problem, especially for high-flying night migrants. Radar gives us increasingly accurate data on the height, density, and speeds of migrants. Tracking equipment on and near Bermuda, for example, has revealed high altitude flights of shorebirds heading southeast, away from land; it has been suggested that these birds later encounter southwesterly winds that make reaching South America easier than would a more direct route.

Techniques other than radar have been used for more precise counts of the number of birds passing one location. A portable ceilometer, casting a beam of light into the sky, is useful in the first hours of night migration when birds are flying low, because the birds passing through the beam can be counted individually. Tracking individuals for any distance, however, is still in a fairly primitive stage — transmitting devices attached to birds do not give signals that can be received as far as the bird may travel in one flight.

A local migration problem deserving further study is the "coastal hiatus" of the coastal plain in southeastern states along the Gulf of Mexico: migrants flying across the Gulf there do not usually land near the coast, but continue further inland, avoiding what seems like perfectly serviceable feeding habitat. This behavior may be a survival from the interglacial periods when water levels were higher and the region was submerged.

Orientation experiments with pigeons and with wild birds continue to reveal the remarkable abilities birds have in finding their way back to a particular location. The mechanisms involved, however, are still not clear; it has been proved that pigeons are acutely sensitive to magnetic impulses and polarized light, but we do not know how they detect and analyze these cues to find their way home.

Experiments that released White-crowned and Golden-crowned

Sparrows wintering in California at varying distances up to 100 miles from home found that more returned from early winter releases than later ones. The highest rates of return were from birds released within 35 miles of home and those released between 60 and 100 miles — those released in between evidently did not or could not use knowledge of the local terrain or any method of true navigation.

The migratory routes and breeding and wintering distributions of many birds are still imperfectly known, although field work in little explored areas is producing much new information. Even in the familiar parts of the world, expansions or contractions of range through manmade or natural causes require continuous documentation — in North America a few new breeding species and many casual vagrants continue to be found. In 1962, three pairs of the normally Old World Little Gull were found breeding in southeastern Ontario; five pairs were found at two Ontario sites in 1971, and in 1975 Little Gulls were discovered nesting in the United States at two Black Tern colonies in Wisconsin. A pair of Black-capped Gnatcatchers nested in Arizona in 1971; the species was previously thought restricted to northwestern Mexico and had never before been recorded in the United States. In 1969 and 1973 the Wood Sandpiper, an Old World species that regularly strays to the Aleutians, was found breeding there on Amchitka Island, but it is suspected that Wood Sandpipers have bred in the Aleutians before, whenever enough individuals happen to be on one island to stimulate courtship displays; they are not actually expanding their New World range as are some recently arrived Alaskan breeders like the Wheatear and Arctic Warbler.

Most of the species recently recorded in North America for the first time have, expectedly, come from the areas nearest other zoological regions. In 1970 a Little Bunting from Siberia landed on a ship in the Chukchi Sea 150 miles north of Alaska. A Common Rose Finch, also from Siberia, was found in the Yukon-Kuskokwim Delta of Alaska in June 1972. In south Texas, flocks of Mexican Crows and increasingly regular sightings of Clay-colored Robins have added these species to the United States list. Among the birds recently found for the first time in Florida is the Bahama Woodstar, a hummingbird. Perhaps the most remarkable recent vagrant to the North American region is a White Tern found on Bermuda in December 1972; it is the first recorded occurrence anywhere in the Atlantic north of the Equator, and is even more surprising because this individual was of a Pacific subspecies — it may have reached the Atlantic while resting or temporarily captive on a ship in the Pacific, or been carried by a series of extraordinary but unrecorded storms.

The discovery of vagrants is exciting, but tells us less about the general distribution of a species than do population studies within its normal range. The many uses of Christmas counts and winter censuses have already been mentioned. Similarly, several kinds of breeding bird surveys are now widely used to measure population changes and centers of abundance; these surveys become more valuable the longer and more often they are done. Some are run like Christmas counts, noting every bird in a prescribed area; some count each bird heard or seen at certain points on a path drawn through particular habitats, and others record nest data for all birds found breeding during the season. Special censuses are taken of waterfowl and many colonial birds; aerial surveys of seabird colonies are used to measure their numbers efficiently with limited disturbance.

The population structure, when it can be determined, is also useful information; it tells us the rate of reproduction, mortality, and how many individuals of each age class are present. For example, Cassin's Auklets, when they breed for the first time, develop a neck pouch that can be used to distinguish all those with breeding experience from nestlings and the nonbreeders, under three years old. In one colony it was found that one- and two-year-old birds made up 30 percent of the population, and new, inexperienced auklets were added to the population at 17 percent each year; the combined mortality rate for fledglings and nonbreeders was estimated at 50 percent, but adult mortality probably matched the population's annual increase of 17 percent.

Some of the data gathered in such studies may seem uselessly specialized, but this research may in fact prove relevant for conservation or other concerns of the non-professional interested in birds. Science, of course, has no particular obligation to provide information that is "relevant," but to the person fascinated by living things and how they interact, all knowledge fits that description. It is hoped that this brief survey of a few topics of current ornithology will show you ways to consider your own observations and will encourage you to investigate further the life of birds in the environment we all share.

Further Reading

The books listed below provide more information on the subjects we have surveyed. Journals, popular or ornithological, have not been included, because back issues are not easily accessible, but *Audubon, Natural History,* and *Scientific American* are the non-professional magazines that most regularly include articles on birds. Addresses for the most important American ornithological journals are given in the Appendix on page 291.

The listings have been arranged by chapter topic, with a brief note at the beginning of each section describing the nature of the books and suggesting other headings to check for related material; since, for example, no book is devoted exclusively to winter behavior, the reader is advised to see the life history books listed for Chapters 8 and 9 and those on migration for Chapter 10.

Where a title is not sufficiently explanatory, it is followed by a brief description; for books published in more than one edition, the most available one is listed.

Chapter 1 — Watching Birds

Books about the sport and technique of watching birds. In addition, for many states and provinces there are now very helpful books detailing the occurrence and distribution of every species found in the region.

Krutch, Joseph W. and Paul S. Eriksson. 1962. A Treasury of Bird Lore. New York: Paul S. Eriksson.
Accounts of interesting observations by ornithologists and bird watchers.

Hickey, Joseph J. 1953. A Guide to Bird Watching. London: Oxford University Press.
Bird watching techniques and suggestions on beginning field studies.

McElroy, Thomas P. Jr. 1974. THE HABITAT GUIDE TO BIRDING. New York: Alfred A. Knopf.
Birds arranged by the habitat in which they are usually found, giving an idea of what to expect where.
Peterson, Roger T. 1957. BIRD WATCHER'S ANTHOLOGY. New York: Harcourt, Brace.
Ornithologists and bird watchers describe adventures and fascinating experiences with birds around the world.
——— *and James Fisher.* 1955. WILD AMERICA. Boston: Houghton Mifflin.
Two eminent ornithologists describe a trip around North America in which they tried to see as many birds as possible in one hundred days.
Pettingill, Olin S. 1951. A GUIDE TO BIRD FINDING EAST OF THE MISSISSIPPI. New York and London: Oxford University Press.
———. 1953. A GUIDE TO BIRD FINDING WEST OF THE MISSISSIPPI. New York and London: Oxford University Press.
Precise directions to outstanding birding localities around the country.
———. 1965. THE BIRD WATCHER'S AMERICA. New York: McGraw-Hill.
Famous ornithologists describe their favorite bird watching places.
Pough, Richard H. 1949. AUDUBON LAND BIRD GUIDE. Garden City, N.Y.: Doubleday.
———. 1949. AUDUBON WATER BIRD GUIDE. Garden City, N.Y.: Doubleday.
———. 1957. AUDUBON WESTERN BIRD GUIDE. Garden City, N.Y.: Doubleday.
More details on habits, food, and nests than other field guides.
Rand, Austin L. 1955. STRAY FEATHERS FROM A BIRD MAN'S DESK. Garden City, N.Y.: Doubleday.
Entertainingly written and illustrated discussion of many unusual and unexpected bird habits, by a leading ornithologist.

Chapter 2 — How and Why Birds are Studied

The history of ornithology, man's relations with birds, how bird behavior is studied, and basic references that discuss all aspects of ornithology in detail.

Allen, Elsa G. 1969. THE HISTORY OF AMERICAN ORNITHOLOGY BEFORE AUDUBON. New York: Russell and Russell.
Biographies of early American naturalists and quotes from their descriptions of the birdlife.

Darling, Lois and Louis Darling. 1962. BIRD. Boston: Houghton Mifflin.
A good general introduction.
Eiffert, Virginia S. 1962. MEN, BIRDS, AND ADVENTURE. New York: Dodd, Mead.
The discovery of American birds, with emphasis on the nineteenth century.
Fisher, James. 1954. A HISTORY OF BIRDS. Boston: Houghton Mifflin.
The history of man's observation of birds, from ancient times to the present; discussions of bird evolution, speciation, populations, and man's present influence on bird numbers.
——— *and Roger T. Peterson.* 1964. THE WORLD OF BIRDS. Garden City, N.Y.: Doubleday.
Well illustrated general introduction, with particularly good distribution maps.
Graham, Frank. 1975. GULLS, A SOCIAL HISTORY. New York: Random House.
Man's impact on the ecology and distribution of gulls in North America, especially the Herring Gull.
Lanyon, Wesley E. 1964. BIOLOGY OF BIRDS. Garden City, N.Y.: The Natural History Press.
An excellent non-technical introduction to the subject.
Nice, Margaret M. 1967. THE WATCHER AT THE NEST. New York: Dover.
How the leading authority on the Song Sparrow pursued her work.
Nicolai, Jurgen. 1974. BIRD LIFE. New York: Putnam.
All aspects of bird life, with particular emphasis on the nesting cycle, and spectacular photographs.
Parmelee, Alice. 1976. ALL THE BIRDS OF THE BIBLE. New Canaan, Conn.: Keats Publishing.
The roles of birds in the Bible, with specific identifications wherever possible.
Pettingill, Olin S. 1970. ORNITHOLOGY IN LABORATORY AND FIELD (4th edition). Minneapolis: Burgess.
Often used in college courses; good anatomical drawings.
Rand, Austin L. 1967. ORNITHOLOGY: AN INTRODUCTION. New York: Norton.
Another good college level text.
Stefferud, Alfred. 1966. BIRDS IN OUR LIVES. Washington, D.C.: U.S. Fish and Wildlife Service.

The influence birds have had on our history and way of life today, and the many ways in which birds are important to mankind.

Stresemann, Erwin. 1975. ORNITHOLOGY: FROM ARISTOTLE TO THE PRESENT. Cambridge and London: Harvard University Press.
A detailed history of the progress of ornithology, by the man many consider the greatest of twentieth century ornithologists.

Thomson, A. Landsborough. 1964. A NEW DICTIONARY OF BIRDS. New York: McGraw-Hill.
Thorough descriptions of every topic in ornithology.

Tinbergen, Niko. 1968. CURIOUS NATURALISTS. Garden City, N.Y.: Doubleday.
A fascinating account of how the behaviorist investigates his subject.

Van Tyne, Josselyn and Andrew J. Berger. 1966. FUNDAMENTALS OF ORNITHOLOGY. New York: Wiley.
An advanced text particularly useful for its extensive glossary and descriptions of each of the world's bird families.

Wallace, George J. and Harold D. Mahan. 1975. AN INTRODUCTION TO ORNITHOLOGY (3rd edition). New York: Macmillan.
A good introductory text.

Welker, Robert H. 1955. BIRDS AND MEN. Cambridge: Harvard University Press.
American birds in science, art, literature, and conservation, 1800–1900.

Chapter 3 — Origin, Evolution, and Speciation

Books on evolution and speciation, and those describing all living families of birds.

Austin, Oliver L. 1961. BIRDS OF THE WORLD. New York: Golden Press.
Large and well illustrated, giving distribution, ecology, and basic life histories for every bird family.

———. 1971. FAMILIES OF BIRDS. New York: Golden Press.
A handy, pocket-sized reference with pictures, descriptions, and characteristics of all families.

Gilliard, E. Thomas. 1958. LIVING BIRDS OF THE WORLD. Garden City, N.Y.: Doubleday.
Excellent photographs and accounts of every bird family.

Mayr, Ernst. 1963. ANIMAL SPECIES AND EVOLUTION. Cambridge: Harvard University Press.
An authoritative analysis of how species evolve.

Nelson, B. 1968. GALAPAGOS: ISLANDS OF BIRDS. New York: Morrow.

The birds that now live on the Galapagos Islands and how they evolved.
Rand, Austin L. 1975. BIRDS OF NORTH AMERICA. New York: Doubleday.
Detailed descriptions of every North American bird family and of some well known species from each. Many good photographs.
Thornton, Ian. 1971. DARWIN'S ISLANDS, A NATURAL HISTORY OF THE GALAPAGOS. Garden City, N.Y.: The Natural History Press. Describes the birds, animals, and plants of the Galapagos, how they reached the islands, and how they evolved.

Chapter 4 — Feathers and Flight

See also the texts listed for Chapter 2 for discussions of feather growth and replacement.
Armstrong, Edward A. 1965. BIRD DISPLAY AND BEHAVIOR. New York: Dover.
Excellent descriptions and photographs showing how different birds use their plumages for display.
Dwight, Jonathan. 1900. THE SEQUENCE OF PLUMAGES AND MOLTS OF THE PASSERINE BIRDS OF NEW YORK. New York: Annals of the New York Academy of Sciences (vol. 13).
Reissued by the Academy in 1975; describes the molts of each species in detail and tells when they take place.
Greenewalt, Crawford H. 1960. HUMMINGBIRDS. Garden City, N.Y.: Doubleday.
Discussions of the unique aspects of hummingbird flight and the iridescence of their feathers, with impressive photographs.
Jameson, William. 1959. THE WANDERING ALBATROSS. New York: Morrow.
Excellent pictures and description of how seabirds use the wind for soaring; a good life history of the Wandering Albatross as well.

Chapter 5 — Food, Feeding, and Digestion

For diet and feeding techniques, see life history studies listed under The Breeding Cycle; for digestion, see the ornithology texts listed for Chapter 2.
Craighead, John J. and Frank C. Craighead Jr. 1969. HAWKS, OWLS AND WILDLIFE. New York: Dover.
A thorough study of the diet of raptors, and the effect they have on wildlife.

Chapter 6 — Anatomy

See the ornithology texts listed for Chapter 2.

Chapter 7 — Voice

See also bird guides for descriptions of each species' vocalizations, ornithology texts for other discussions of sound production, and life histories for the behavior associated with vocalizations.

Armstrong, Edward A. 1973. A STUDY OF BIRD SONG. New York: Dover.
A thorough review of all aspects of bird song.
Greenewalt, Crawford H. 1968. BIRD SONG: ACOUSTICS AND PHYSIOLOGY. Washington, D.C.: Smithsonian Institution Press.
An analysis of how sounds are produced.
Thielcke, G. A. 1976. BIRD SOUNDS. Ann Arbor: University of Michigan Press.
A good survey of both vocal and mechanically produced sounds.

Chapters 8 and 9 — The Breeding Cycle

All detailed studies of families and individual species are listed here. In addition to information on the breeding cycle, these studies naturally discuss migration, winter habits, diet, vocalizations, and other aspects of bird behavior.

Bent, Arthur C. 1961–1976. LIFE HISTORIES OF NORTH AMERICAN BIRDS. New York: Dover.
Thorough biographies of every native species, in a many-volumed series.
Brown, Leslie and Dean Amadon. 1968. EAGLES, HAWKS AND FALCONS OF THE WORLD. New York: McGraw-Hill.
Delacour, Jean. 1954. THE WATERFOWL OF THE WORLD. London: Country Life.
——— *and Dean Amadon.* 1973. CURASSOWS AND RELATED BIRDS. New York: American Museum of Natural History.
Fisher, James. 1952. THE FULMAR. London: Collins.
——— *and Robert M. Lockley.* 1954. SEA-BIRDS. Boston: Houghton Mifflin.
Natural history of seabirds of the North Atlantic.
Forshaw, Joseph M. 1973. PARROTS OF THE WORLD. Melbourne: Lansdowne.
Gilliard, E. Thomas. 1969. BIRDS OF PARADISE AND BOWER BIRDS. Garden City, N.Y.: Natural History Press.
Griscom, Ludlow and Alexander Sprunt. 1957. THE WARBLERS OF AMERICA. New York: Devin-Adair.
Grossman, Mary L. and John Hamlet. 1964. BIRDS OF PREY OF THE WORLD. New York: Clarkson N. Potter.
Harrison, Hal H. 1975. A FIELD GUIDE TO BIRDS' NESTS IN THE

UNITED STATES EAST OF THE MISSISSIPPI. Boston: Houghton Mifflin.
Color photographs of the nests and eggs of nearly all 285 species described.

Hochbaum, H. Albert. 1955. THE CANVASBACK ON A PRAIRIE MARSH. Harrisburg, Pa.: Stackpole.

Johnsgard, Paul A. 1973. GROUSE AND QUAILS OF NORTH AMERICA. Lincoln: University of Nebraska Press.

———. 1975. WATERFOWL OF NORTH AMERICA. Bloomington and London: Indiana University Press.

Lack, David. 1968. ECOLOGICAL ADAPTATIONS FOR BREEDING IN BIRDS. London: Methuen.
A thought provoking analysis of the effects of the environment on nest site, clutch size, and growth rate.

———. 1970. THE LIFE OF THE ROBIN. London: Collins.
A classic life history of the European Robin.

Lockley, Ronald M. 1947. SHEARWATERS. New York: Devin-Adair.
Life history of the Manx Shearwater.

———. 1953. PUFFINS. London: J. M. Dent.
Life history of the Atlantic Puffin.

Murphy, Robert C. 1936. OCEANIC BIRDS OF SOUTH AMERICA. New York: Macmillan.
Ecology and life histories of seabirds, including many found in North America.

Palmer, Ralph S. 1962. HANDBOOK OF NORTH AMERICAN BIRDS (vol. 1). New Haven and London: Yale University Press.

———. 1976. HANDBOOK OF NORTH AMERICAN BIRDS (vol. 2). New Haven and London: Yale University Press.
Volume 1 covers loons through flamingos, Volume 2 waterfowl; thorough accounts of plumage, life history, and distribution.

Simpson, George G. 1976. PENGUINS: PAST AND PRESENT, HERE AND THERE. New Haven and London: Yale University Press.

Skutch, Alexander F. 1973. THE LIFE OF THE HUMMINGBIRD. New York: Crown.
Survey of the entire family.

———. 1976. PARENT BIRDS AND THEIR YOUNG. Austin and London: University of Texas Press.
A thorough review of all aspects of the breeding cycle, with much information on tropical species.

Smith, L. H. 1968. THE LYREBIRD. Melbourne: Lansdowne.

Stout, Gardner D. (ed.) 1967. THE SHOREBIRDS OF NORTH AMERICA. New York: Viking.

Tinbergen, Niko. 1960. The Herring Gull's World. New York: Basic Books.
An outstanding behavior study of Herring Gulls in the breeding season.
Tuck, Leslie M. 1960. The Murres. Ottowa: Canadian Wildlife Service. The evolution, distribution, and biology of the Common and Thick-billed Murres, and their economic value to man.
Walkinshaw, Lawrence H. 1973. Cranes of the World. New York: Winchester.

Chapter 10 — Migration

Listed here are books dealing exclusively with migration and orientation.

Dorst, Jean. 1962. The Migration of Birds. Boston: Houghton Mifflin.
The most complete review of all aspects of bird migration.
Eastwood, Eric. 1967. Radar Ornithology. London: Methuen.
The fascinating and sometimes surprising information learned from tracking birds with radar equipment.
Griffen, Donald R. 1964. Bird Migration. Garden City, N.Y.: Natural History Press.
A non-technical survey of major aspects of bird migration.
Heintzelman, Donald S. 1975. Autumn Hawk Flights: The Migrations in Eastern North America. New Brunswick, N.J.: Rutgers University Press.
A thorough discussion of autumn hawk migration in eastern North America — identification, lookouts, flying techniques, routes, and the weather conditions that produce large flights.
Lockley, Ronald M. 1967. Animal Navigation. New York: Hart.
Navigation techniques throughout the animal world, making interesting comparisons with those of birds.

Chapter 11 — Winter Habits

See books listed under The Breeding Cycle and under Migration.

Chapter 12 — Distribution

In addition to those listed here, all general books on ornithology discuss distribution, and the bird guides now available for many parts of the world illustrate most species found in their region.

Amadon, Dean. 1966. Birds Around the World. Garden City, N.Y.: Natural History Press.
A good survey of all aspects of distribution.

Darlington, Philip J. 1957. ZOOGEOGRAPHY: THE GEOGRAPHICAL DISTRIBUTION OF ANIMALS. New York: Wiley.
An advanced and detailed discussion of the distribution of all animals; especially interesting if you want to compare the process and geography of bird distribution with that of other types of animals.
Dunning, John S. 1970. PORTRAITS OF TROPICAL BIRDS. Wynnewood, Pa.: Livingston.
Handsome photographs of many rarely illustrated Neotropical birds.
Fry, C. Hilary and J. J. M. Flegg (eds.). 1974. THE WORLD ATLAS OF BIRDS. London: Mitchell Beazley.
A well illustrated survey of bird ecology based on distribution.
Udvardy, Miklos D. F. 1969. DYNAMIC ZOOGEOGRAPHY, WITH SPECIAL REFERENCE TO LAND ANIMALS. New York: Van Nostrand.
Another good book for comparisons of all types of animal distribution.

Chapter 13 — Conservation

See also the books listed for Chapter 2 on the relations of men and birds.

Broun, Maurice. 1949. HAWKS ALOFT: THE STORY OF HAWK MOUNTAIN. New York: Dodd, Mead.
Describes the spectacular hawk flight visible each fall from Hawk Mountain, Pennsylvania, and tells how the site became a sanctuary rather than a shooting ground for hawks.
Greenway, James C. Jr. 1967. EXTINCT AND VANISHING BIRDS OF THE WORLD. New York: Dover.
The most authoritative and complete reference on the subject.
Harwood, Michael. 1973. THE VIEW FROM HAWK MOUNTAIN. New York: Scribner's.
An update on Broun, with descriptions of how pesticides affect the reproductive abilities of hawks.
MacMillan, I. 1968. MAN AND THE CALIFORNIA CONDOR. New York: Dutton.
Matthiessen, Peter. 1959. WILDLIFE IN AMERICA. New York: Viking.
Man's role in the decline and extinction of bird and animal species in North America.
Mayfield, Harold F. 1960. THE KIRTLAND'S WARBLER. Bloomfield Hills, Mich.: Cranbrook Institute of Science.
McNulty, Faith. 1966. THE WHOOPING CRANE. New York: Dutton.
The status of the Whooping Crane in the wild and in captivity up to 1966.

Zimmerman, David R. 1975. TO SAVE A BIRD IN PERIL. New York: Coward, McCann.
Innovative techniques in the conservation of endangered birds.

Chapter 14 — Attracting and Caring for Birds

Check also publications of local garden clubs, nature organizations, etc.

Arbib, Robert S. and Tony Soper. 1971. THE HUNGRY BIRD BOOK. New York: Taplinger.
Attracting and caring for birds east of the Great Plains.

Davison, V.E. 1967. ATTRACTING BIRDS FROM THE PRAIRIES TO THE ATLANTIC. New York: Crowell.

Dennis, John V. 1975. A COMPLETE GUIDE TO BIRD FEEDING. New York: Knopf.

Grant, Karen A. and Verne Grant. 1968. HUMMINGBIRDS AND THEIR FLOWERS. New York: Columbia University Press.
The ecological relationships of hummingbirds and the flowers on which they feed.

Terres, John K. 1968. SONGBIRDS IN YOUR GARDEN. New York: Crowell.
Useful throughout North America.

Chapter 15 — Ornithology Today

Books listed here discuss a few of the current major interests of ornithology, but, of course, the topics of books listed elsewhere continue to be subjects of research.

Farner, David S. and James R. King (eds.) 1970–1975. AVIAN BIOLOGY (vols. 1–5). New York: Academic Press.
Detailed, professional reviews of current ideas in systematics, ecology, anatomy, and physiology.

Lack, David. 1966. POPULATION STUDIES OF BIRDS. London: Oxford University Press.

———. 1971. ECOLOGICAL ISOLATION IN BIRDS. Cambridge and London: Harvard University Press.
A review of the subtle ways birds reduce competition between species.

Snow, David W. 1976. THE WEB OF ADAPTATION: BIRD STUDIES IN THE AMERICAN TROPICS. New York: Quadrangle/New York Times Book Co.
Problems in tropical bird ecology.

Appendix: Ornithological and Conservation Organizations

Following are addresses of the major American ornithological organizations, with the journals they publish, as described in Chapter 2, and the conservation organizations mentioned in Chapter 13.

Ornithological Organizations and their Journals

American Birds. NATIONAL AUDUBON SOCIETY. 950 Third Avenue, New York, N.Y. 10022.

The Auk. AMERICAN ORNITHOLOGISTS' UNION. National Museum of Natural History, Smithsonian Institution, Washington, D.C. 20560.

Bird-Banding. NORTHEASTERN BIRD-BANDING ASSOCIATION.* c/o Manomet Bird Observatory, Post Office Box O, Manomet, Mass. 02345.

Birding. AMERICAN BIRDING ASSOCIATION. Box 4335, Austin, Texas 78765.

The Condor. COOPER ORNITHOLOGICAL SOCIETY. Department of Zoology, University of California, Los Angeles, Calif. 90024.

The Wilson Bulletin. WILSON ORNITHOLOGICAL SOCIETY. The Museum of Zoology, University of Michigan, Ann Arbor, Mich. 48104.

* Publishes *Bird-Banding* for all the regional banding associations.

Conservation Organizations

International Council for Bird Preservation. c/o Dr. S. Dillon Ripley, Smithsonian Institution, Washington, D.C. 20560.

National Audubon Society. 950 Third Avenue, New York, N.Y. 10022.

World Wildlife Fund. 1319 Eighteenth Street, N.W., Washington, D.C. 20036.

Index